HOUSE BURNED.
Building Destroyed on Sunday Night.
kept Port Colden in a ferment for several hours Sunday evening in the power-house at Plane No. 6, ...ace. The building and the machinery were entirely ... The blaze was discovered ...clock shortly after a "hog" ...attached to a heavy train ...and the natural supposition ...a spark from the engine ...use of the conflagration. ...ne at that point is a double ...of three of that kind on the ...anal. For several years, how... ...y one track had been in use,

Co., Lessees Morris Canal.

pg. 11 missing

The Morris Canal & Banking Co.

Jersey City Mar 30th 1867

Morris Canal & Banking Co.

JERSEY CITY

RECEIVABLE FOR CANAL TOLLS.

STATE OF NEW JERSEY

TWO

Twelve Months after date the Morris Canal & Banking Co. Will pay Two Dollars to Bearer with interest

JERSEY CITY Aug 12 1841

Cash. Pres.

The officers of the Morris Canal passed through Hackettstown yesterday on an inspecting tour. The boat was drawn by two fine mules and the accomodations on the "vessel" were evidently first-class.

June 6, 1874

10 Shares

Morris Canal & Banking Company.
OF 1844

General Office of Morris Canal.

Phillipsburg, N. J. Jan 16 1894

...ived of Thos Paulis per Margot

Dollars,

Lehigh Valley Rail Road Co., Lessees,
per

President.

Tales the Boatmen Told

edited with an introduction by

James Lee

1977

Canal Press Incorporated

Exton, Pennsylvania

I respectfully dedicate this book to Isabelle Lenstrohm Mann.

She has been very helpful to me over the years, through her recollections, knowledge and love of the Morris Canal.

Tales the Boatmen Told

 For information, contact Canal Press Incorporated, Box E, Exton, Pa. 19341
Library of Congress catalog card number: 77-089270

ISBN: 0-916838-08-0

Printed in the United States of America

First Printing August 1977

TABLE OF CONTENTS

Famous Morris Canal Mules
(of the early 1890's)

Faithful mules and horses,
They toiled so hard and long,
Pulling heavy canal boats,
While boatboys walked along.

Boatmen shouted to their mules
The French word "Petit-Whoa!"
For boats coming out of locks
Had to be pulled, just so.

Gunderman's educated mules
Had no driver day or night,
Andy claimed the mules could read
And without a doubt could write.

Billy Blizzard had his "Buckskins",
Dan Bush's "Bays" were reddish brown,
No finer teams were ever seen
When these two passed through town.

Pat Smith drove "Kicking Kate"
Bill Johnson had "Mighty Blacks"
Beecher Dagon called his "Hop-Toads"
Webb Chewman's were "Chubby Jacks."

Dan Mowder drove his "Blue Mule"
He'd call to one and all.
Those canallers had their stories,
And some were mighty tall.

Yes, those good old mules are gone now,
But I can hear them even today.
The clump of their shoes on the tow-path,
Their snorts and affectionate bray.

[Author's Note: This poem, written by Lorraine Willever of Belvidere, New Jersey, was made possible by material supplied by Charles Matlock Hummer, a former boatman who wrote the poem, "Famous Tiller Sharks."]

PREFACE

Jim Lee, a railroader by trade, bought the plane tender's house at Plane Number 9 West of the Morris Canal when he came back from the Second World War. For thirty years he has been quietly carrying on historical activities about the canal, of a scope that would rival a well-staffed and well-funded organization. In a small and rather old-fashioned office in his home he has assembled a vast collection of pictures, publications, postcards, manuscripts, maps, artifacts and tape recordings, along with photographic and recording equipment. He has become an expert on the history of the canal, and his lecture-slide presentation is much in demand. He has undertaken the extremely difficult job of restoring the turbine that powered the inclined plane on his property, which was buried under tons of rock when the canal was abandoned in the 1920's. And he has published a pictorial history of the canal, the first of his books on the subject.

I was introduced to Jim Lee a few years ago by former Senator William E. Schluter, who was then vice-chairman of the New Jersey Historical Commission. One result of our visit was a modest grant-in-aid to Jim for an oral history of the Morris Canal, through interviews with some fifteen or twenty people who had lived and worked on the canal. The generation that had known the canal in its heyday in the nineteenth century was long gone, of course. But the recollections of these few survivors could at least help us preserve the culture—the experiences, the lore, the music—of canal life in its last three decades.

The project was finished last year, and one Sunday in June our associate director, Dick Waldron, and I went up to visit with Jim. We sat around for about an hour and listened to the tapes of the interviews. Time receded and we were back in an era of long boats laden with coal ore, of inclined planes and locks, colorful captains and their families, mules and their drivers (mostly children), plane tenders and lock tenders, Sunday school and baptisms in the canal, and a whole world of experiences encountered on the once ordinary, now fabulous journey between the terminals of Phillipsburg on the Delaware and Jersey City on the Hudson. And the songs. To hear the three or four canal boat songs sung by old voices with the freshness of childhood is an experience in itself.

The public has already had the fruits of this project, through the film, "Famous Tiller Sharks," produced and shown by the state's Public Broadcasting Authority. And the Canal Society of New Jersey has featured the oral history in its public program, "Tales of the Morris Canal." But the most important thing is that the tape recordings exist—that one man, working for many years with little publicity and no material reward, has given the people of New Jersey a part of their history that would otherwise have soon faded from memory forever.

Our Commission's part in this has been modest enough. Nonetheless, it is one of our functions to find out about people like Jim Lee and to give them whatever assistance we can. We believe his oral history of the Morris Canal is one of the best projects we have supported.

In the present volume the recollections and the lore of the old-timers have, for the first time, been transcribed and edited to produce an important historical source book. It will be a unique tool for the scholars. It will be cherished by men and women whose spirits are moved by a sense of past life. Like the canal that was, that past flows through the lives of all of us.

Trenton
April 1977

Bernard Bush
Executive Director
New Jersey Historical Commission

FOREWORD

It has been a good many years since first I was introduced to the Morris Canal by a homemade raft and a long pole. The location of this introduction was the boatyard at Phillipsburg, New Jersey. There in the shadow of Mt. Parnassus, a large rock formation which looms high above Port Delaware, I found my land of make-believe.

To a boy growing up, it was the next best thing to heaven. I didn't know too much about heaven, except that you went there when you died, if you were good. But I did know about the coal chutes and the canal. What a wonderful place to play! Climbing the mountain, baking potatoes in a fire, rafting on the canal, fishing for sunnies, throwing stones at water snakes, ice skating in season, but no swimming. There were too many leeches, snakes, and too much seaweed.

Little did I know that this early introduction was to become first—a playground, second—a hobby, and now—an obsession.

I don't believe there was another canal in all the world that combined the natural beauty of God's creation with the progressive spirit and ingenuity of man.

Most canals followed rivers or streams and had their source of water nearby, but not so the Morris. It left a good supply (Delaware River) and struck out across the rolling hills and mountains of northern New Jersey. By utilizing locks and inclined planes, it climbed 914 feet above sea level on its route from Phillipsburg to Jersey City.

This engineering feat made the Morris Canal a marvel to behold, but much more than its machinery and other physical aspects were the people.

In 1974 I received a grant from the New Jersey Historical Commission which enabled me to conduct an oral history/folklore project among the very few remaining people who had a direct relationship with the Morris Canal. It was the last roundup of primary information from people who boated, worked, played, fell in, or just lived along this wonderful waterway.

The folklore, stories, and songs that you will encounter in this book, *Tales The Boatmen Told,* are only a few of what must have been many. Let us not lament the ones that have been lost, but rejoice in the ones that have been saved.

The Morris Canal was built to be a commercial carrier of goods, both raw and manufactured, but it was much more than a water highway. It became a way of life for thousands of people over the almost one hundred years of its existence.

People will not remember this canal for the shoddy financial manipulations of its bankers, or the low returns paid to its investors. But rather they will remember the challenge of its construction, the growth and prosperity it brought to the State of New Jersey, and the joys and sorrows that came to the people who were canalers.

This book has been published to provide insights to the scholars who might want to know what everyday life was like on the canal, to bring the emotions and thoughts of the canalers to the hearts and minds of the readers, and to make sure that this rich heritage is kept safe for future generations.

ACKNOWLEDGEMENTS

I wish to acknowledge that I am indebted to so many for all the stories I have gained about the Morris Canal. Many of these people have long since died. But the information and folklore passed on to me is very much alive, and will, I hope, live long after I'm gone.

I want to acknowledge with thanks, the grant from the New Jersey Historical Commission which enabled me to conduct the oral history/folklore interviews with former canal people. The fact that four of these people have died, and two others are in serious physical condition since I undertook this project, points out that this is probably the last roundup of primary information about canal people.

To Harry Rinker, I owe a very special thank-you. It was his constant encouragement, expertise and patience that made this book possible.

I want to acknowledge my satisfaction with the illustrations by Kenneth Knauer. These have been ably created from photographs and stories I supplied to him.

I also want to give special thanks to my friend, Ronald Wynkoop. He has never failed me when I needed a copy of a photograph on short notice, or in any other photographical problems, or special historical research.

The people that I interviewed via the oral history/folklore project will always have a special place in my heart. As you read this book, you will know, and I hope, love them as I do.

In addition to the people mentioned in this book, I want to acknowledge with a special thank-you, the following:

Mrs. Gina Bellini
Mr. William Black
Mr. Wayne Condon
Mr. Warren Crater
Mr. Pfrommer Dailey
Mrs. Joan Fling
Mr. Edward Francis
Mrs. Gladys Fuehrer
Mrs. E. Helen Getter
Mr. & Mrs. Paul Hummer
Mr. Russell Harding
Mrs. Rose McCann
Mr. Thomas Roe
Mr. Martin Rush
Mr. Clayton Smith
Mrs. Edith Steel
Mr. William Stone
Mr. John Vanderbilt
Mrs. Carrie Winters
Mr. James Woodruff

ALSO

Canal Society of New York State
Canal Society of New Jersey
Dover Public Library
Historical Society of Schuylkill County
Lake Hopatcong Historical Society
Morristown Public Library
Pennsylvania Canal Society
Rutgers University Library

ISABELLE LENSTROHN MANN

Age 78 in 1976. Born April 1898 on a canalboat near Port Colden. Daughter of canalboat Captain Peter Lenstrohm and granddaughter of Captain Martin Van Sickle. Now living in Phillipsburg.

Chapter One

Isabelle Lenstrohm Mann

"I was born and raised on the canal . . . Oh yes, I was born between Washington and Port Colden on a Morris Canal boat."

The following is a taped conversation of Mr. James Lee interviewing Mrs. Chester Mann:

What is your full name?
Isabelle Lenstrohm Mann.

And how old are you, Mrs. Mann?
76.

And do you have any objection to me taping this?
No, I'd love to have people hear what I have to say.

What was your connection with the Morris Canal?
I was born and raised on the canal.

When you say born, do you mean you were actually born on a Morris Canal boat?
Oh, yes. I was born between Washington and Port Colden on a Morris Canal boat.

And what year would that be?
1898.

And was there a doctor in attendance?
No. There was a lady friend of my mother's, a Mrs. Devins helped my mother that night.

She was a midwife?
She was a midwife. Women like that were called midwives.

And did her family work on the canal, too?
Yes. Her husband and her; they had a couple of boats on the canal.

And where did this happen?

Between Washington Port and Port Colden.

Had it been planned that way?

No. My mother was trying to get home to Dr. Barber in Phillipsburg.

Was your home in Phillipsburg?

In Phillipsburg, yes.

Do you have any brothers and sisters?

Yes, there was eight of us: four boys and four girls. Four sisters and four brothers.

Were any of those born on the canal boat?

No. I was the only one.

And how many of your brothers and sisters actually went with your mother and father?

My oldest brother, John and Amos went after me.

But none of the girls?

None of the girls. My sisters were mostly home. They were younger and stayed home.

How many years old were you when you last remember accompanying your mother and father on the canal?

Fourteen years old when I last went on, as I can remember.

What did you do when you left the canal?

I worked in the silk mill in Phillipsburg.

Did your father continue to boat on the canal?

He quit boating and worked on the scow with the maintenance crew. He stayed on that until the canal broke up.

Did you have any other relatives working on the canal?

Yes, my grandmother and grandfather—Ruth and Martin VanSickle were captains on the canal.

That was your grandfather and grandmother on whose side?

My mother's side.

And did your father have any relations?

Yes, he had relation but they didn't work on the canal. They lived in Jersey City. And Uncle John and Aunt Kate VanSickle worked on the canal. And then my Uncle Amos

The digger crew maintained the five foot depth of the canal. Straddling the gear wheel is Amos Van Syckle, Mrs. Mann's uncle.

worked on the digger. It was called the mud digger. And my cousin Harry worked on the mud digger and worked on the payboat, Harry did.

Do you remember loading coal in Phillipsburg?

Oh yes, loading and unloading.

Was that quite a highlight in your life, to watch the coal come into the boat?

Oh yes, we used to love to watch it come in. We would get awful dirty from the dust and we would have to keep it off the place where we walked on the boat.

How long would it take to load the boat at Phillipsburg?

Oh, it used to take about three hours sometimes–two to three hours to load it.

I suppose you had to load it a little bit and then pull up a little bit so the coal would be evenly spaced?

Yes, they disconnected the boat at the hinges and first they would load one side and then they'd load the other and then pull it together and hinge it.

They would hinge it up. Well then, there was two boats in one. The front section and the back section.

Yes.

Coal was transferred from railroad hopper cars to canalboats at the Port Delaware chutes near Phillipsburg.

Boats carried many commodities in addition to coal, such as wood. Location: near Denville, probably Savage Road.

Where did you generally take this coal?

We took it all over. We took it to Stewartsville to Mr. Stone and to Mr. Dowling and then we took it on further. I forget the man's name in Washington; we used to unload there. There was one in Hackettstown and the one place that I can remember good is the foundry at Rockaway and Powerville.

Did you go to Paterson or Jersey City?

I did when I was smaller, but when I had to drive the mules we didn't go that far.

When you had to drive mules–how do you drive a mule?

Well, just walk behind them and guide them; talk to them and when we would get to a spot where they had to pull hard, then we'd, then we'd get up and get ahold of them in front where the bit is and lead them.

Were they generally pretty gentle?

Yes, the mules were more gentle than the horses.

If you weren't there to guide them, what would they do?

Well, they would stop and eat. If they seen something they wanted to eat, they'd stop and eat.

Well, what other commodity did you carry besides coal?

Well, we carried lumber, sawdust, ice and manure. And I can remember once bringing nails from Boonton in kegs, and heavy timber that we would use for the bottom of the locks. The floors of the locks all had timber on them and so did the foot and top of the plane have heavy timber. They used to leave the heavy planks off at each lock or plane that might need them. Then the rest we'd take to the boat yard. Sometimes it was ties and sometimes heavier and longer timber than ties.

How about sawdust. Where would you take the sawdust?

The sawdust we would take to the icehouses that were located along the canal. There was a lot of icehouses along the canal. The different ice companies had them. I can remember in Phillipsburg it was Hagerty's.

Where did you get the sawdust from?

We would get the sawdust from the sawmills along the canal. There were several sawmills along the canal. I can remember one in particularly in New Village, behind New Village.

You mentioned manure—where did you take the manure?

We'd take it to Hackettstown and then unload it there. They had a big building along the canal. I think they took it to Great Meadows and up through Shades of Death at that time.

Did you ever bring back any odd loads, anything that was out of the ordinary? I think you mentioned poison at one time.

Oh yes, oh yes, "rough-on-rats" from Jersey City.

Where would you get that from?

Jersey City. From the Rough-on-Rats Works in Jersey City.

Why was it rough on rats?

LADIES, IT'S JUST LOVELY

Send 10 Cents to E. S. WELLS, Jersey City, N. J., and receive by return mail a beautiful "ROUGH ON RATS" IRON HOLDER. Its splendid.

Also, 15 Cents for a New Chromo, in seven colors, 13x21 inches, elegant for any room or office, entitled "HOUSEHOLD TROUBLES." Best thing out.

And 10 Cents for a set of large size SCRAP BOOK CARDS in colors; amusing, instructive, beautiful.

And 35 Cents for Song and Chorus of "ROUGH ON RATS." This is immense. Just out. E ybody crazy for it.

An 'ents for "THE SEASIDE SIBYL; OR LEAVES OF DESTINY." A fortune teller in verse Filled with comic illustrations.

ALL TOGETHER, 40 CENTS.

"ROUGH ON RATS."

Clears out rats, mice, roaches, flies, ants, bed-bugs, skunks, chipmunks, gophers, beetles. 15c. Druggists.

"BUCHU-PAIBA."

Quick, complete cure, all annoying Kidney, Bladder and Urinary Diseases. $1. Druggists.

CATARRH OF THE BLADDER.

Stinging irritation, inflammation, all Kidney and Urinary Complaints, cured by "Buchu-paiba." $1.

FLIES AND BUGS.

Flies, roaches, ants, bed-bugs, rats, mice, gophers, chipmunks, cleared out by "Rough on Rats." 15c.

This "Rough-on-Rats" was suppose to kill rats and mice and things. They had it done up in little boxes and put in wooden cartons. They didn't have paper cartons at that time but they were all heavy wooden cartons so you couldn't get in them. Children couldn't get in them. Children couldn't get in them because it was poison.

How many hours a day did your boat generally work?

Well, usually sun-up to sun-down. Sometimes it was dark when we stopped. If we wanted to get someplace in a hurry (if we weren't in a hurry we would stop a little early), but it would be from around half past six in the morning till around half past seven at night and sometimes longer if we was in a hurry and wanted to get somewhere.

Can you bring to your mind any interesting happenings that you always remember on the canal? Something out of the ordinary?

Well, baptism and different things like that they used to do along the canal. The Baptists used to baptize the people right in the canal. I saw that happen.

Did you ever see anyone drown in the canal?

Oh yes! I saw one of the lock tenders children fall between the boat and the lock.

Was he drowned or was he crushed?

Crushed, and that was at Fluke's lock.

And where would you say that would be near?

Stanhope.

How about your boat number. Do you remember any numbers?

Yes, 713 and I think the last one was 736. I kind of forget, but I am pretty sure it was 736. The 713 we had a long time.

Did they try to keep the same boat?

They tried to keep the same boat for good luck.

Did your father have any pet names for his boats?

Yes. The first boat I can remember that he had was "Rachael". He called it "Rachael" and whenever it would get stuck or anything he would say "Damn you, Rachael, what's the matter with you?"

Did the children ever give your father a hard time?

Oh yes, especially when they were swimming. They would get ahold of the boat rudder and hang on to it and then he couldn't move it.

Even out of water swimmers were attracted to the canalboats. Location: Plane 2 West, near Stanhope.

So he used to get mad and turn around and he was a tobacco chewer and chew tobacco and he would wait till he would get a whole mouth full of juice and then he would squirt it right down in their face and in their eyes. Then they would leave go of the boat and leave go of the rudder and we would go on. So then that would stop them and they very seldom would do it after that, when they found out who he was they would just stay away.

Did anyone ever play tricks on your father or any of the boatmen that you know of? Did they do anything, not to damage it but to harass him in any way?

Oh yes, they called him names and they used to holler:

You rusty canaler
You'll never get rich
You'll die in a cabin
You son-of-a-----

Did they ever throw tomatoes at you?

Oh yeah, they did that when we were small and sat on the cabins. A whole lot of that was in Newark on Market Street. They used to throw their junk and stuff right down on top of our boats.

How many days a week did you boat?

Well, we boated practically every day if we was in a hurry, but we used to like to tie up on Sunday. It was a necessity to tie up on Sunday.

They didn't operate the locks or planes on Sunday?

No. Unless we got to one of the planes and got through and then went on but we couldn't lock through.

Did you go to church along the way?

Oh, yes. I never went to church out of Phillipsburg. But in Phillipsburg we always got off. My aunt used to get me down on Mill Street down by the culvert. My dad used to put me on the hook pole and swing me over to the towpath and she'd catch me. And then from there on to her place I'd go to get cleaned up and go to church. We went to St. Luke's Episcopal Church there in Phillipsburg. Reverend Martin was the minister when I was little.

Did you ever do any singing on the canal on Sunday?

Oh yes, all kinds of singing. And they would have little parties on Sunday and play music. They played the accordian, the violin and the Jew's harp and on the empty boats they used to make their own music, Jim. They would have a couple of bottles–I imagine they were beer bottles–and they would fasten wire to them somehow or other on the bottom of the boat and they played music with them with their fingers and with a stick. I don't know what they called that but they used to do that.

Was your father musically inclined?

Yes, he played the accordian, harmonica, Jew's harp and the violin–they called it a fiddle.

How about the brothers, were they?

Oh yes. They were all musically inclined. They played banjo, violin, guitar. My mother, too. She played the accordian and the fiddle and the Jew's harp.

Did they have any dances at different times?

Oh yes, if we were tied up at Port Delaware or any good place where there was a good place to dance—an old barn or something—why, they would have square dances.

Do you remember a minister going down to Port Delaware talking to you? Do you remember who it was?

Oh yes, Reverend Williston. He used to come down and give the children little cards with a verse on the back and his wife used to take care of a lot of the people on the canal. She was a doctor.

You said there was a verse on the back. What was on the front?

Pictures of Christ and angels and things. I can remember them real well.

Did he take up a collection?

Never took up a collection. I don't remember his ever taking up a collection. Just coming and preaching a sermon for us and singing hymns.

No music though?

No, no music.

Did your father make money on the canal?

Oh yes, as far as I know we always got along.

Did you always have enough to eat?

Oh yes.

Did you ever supplement your income by trading coal for groceries?

Not for groceries so much as along the canal we used to trade with the farmers. Apples, potatoes and onions and things like that, and corn that they wanted to give away. We never had to throw our coal to get that stuff—they gave it away. We would trade coal for it.

Then were you able to save enough coal to last you through the winter, too?

We never saved for ourselves.

"Hunks-a-go pudding and pieces of pie?" Oh yes, very well. Shall I sing it for ya?

"Yeah, if you feel up to it, I would like to hear it."

Hunks-a-go pudding and pieces of pie
Me Mother gave me when I was a boy
And if you don't believe it, just drop in and see—
The hunks-a-go pudding me mother gave me.

That was a little ditty they sang?
Yeah, that was a little ditty we sang on the canal. They played it to music.

What was hunks-a-go pudding?
Hunks-a-go pudding was a pudding they made. It was made if they had a beef roast and they would make a batter of eggs, milk and flour and put it in this fat that was left from the beef roast and put it in the frying pan, put the lid over it and put it on top of the stove. We didn't have any ovens on the boat. A few people had stoves with ovens. Later on my mother had a stove with a little oven. But usually it was out by the hinges, the stove was, and we cooked our food and stuff out there.

Do you remember any other four-line ditties you used to sing?
Oh yes:

You take the hatchet
And I'll take the saw
We'll saw the legs off
Me mother-in-law.

There wasn't much to it but they sang them.

The boys as well as the girls sang?
Oh yeah. Everybody sang.

I guess sometimes you were too tired to sing, though, weren't you?
Well, yeah, after walking all day. But they sang and whistled all day a lot along the boat while we were driving mules.

When you would drive these mules and quit for the day, what would be the procedure—what would you do with the mules?

Well, we would have to take and curry them, clean them because they would be sweaty and the hair would be loose. We had to curry them good around the collar where the collar fitted around the neck.

In other words you would brush them?

Yeah.

Then what would you do with them?

And then usually along the plane and sometimes the locks had them, too—they would have a place to put them each night. We would have to bed them down with straw, clean the stables, and bed them down with straw.

And then you had to pay a little for the straw? You carried your own feed with you?

Yeah, we had our own feed. We had it by the hinges, there was a big box.

And when you got the mules taken care of, did you get a bite to eat then?

Then we got supper and sat and read or played games. If the boat was empty, we had swings in the empty part. We played marbles, and other games such as dominoes, checkers or even played with a cord on our hands to draw things.

What did you make out of the cord?

Well, we made "the cat-in-the-cradle" and different things like that. And we took a spool and put four little nails on it and wove a rope and took this rope and played with it during the days. We also played with our little dolls that we made out of clothes pegs and corn husks and we had apple-head dolls that we made.

How did you make them?

Well, they dried the apples somehow—our mothers and grandmothers made them for us. They dried the apples and fixed them and put them on sticks and they had wooden bodies. They fixed dolls and things like that for us to play with. And the boys they usually played with the marbles and things, or something like that. I don't know what else to tell you about what the boys done.

Where did you sleep when there was a mixed family like boys and girls?

Well, we had bunks, but usually the boys slept on the floor and the girls slept in the bunks if we were with my mother and father. My mother and father—they were always on the lower bunks; we had the top ones. They were big enough for two to sleep in.

How were your meals on the canal boats—were they pretty good meals, generally?

Oh yes.

Were they adequate—did you ever go hungry?

I don't remember that we ever went hungry because we made soup called pork float—that was a regular canaler's soup. Everybody made that and it was made from salt pork and our meat was almost always salted because you couldn't keep it unless it was salted. There was no refrigeration at that time. So we used the salted pork, and the pork float was made with pork. And we used to saute the pork a little bit before we put it in the soup and the potatoes, onions and tomatoes. Some people like corn in it but we never did. We just put potatoes, tomatoes and onions and it was really delicious. Pepper, salt

Cooking conditions were primitive and the meals had a certain sameness. Boatmen were lucky when their wives went along.

and a little parsley and a sweet marjoram—whichever you wanted in it. Then there was bean soups we made and corned beef and cabbage.

How about Sunday? Was Sunday a special day for eating? Was there something you ate more on Sunday?

Yes, we always had beef roast on Sunday because you didn't get too much of that.

How about salt mackerel—was that a favorite?

Oh yes, they would boil that off. That was mostly a breakfast meal. That was an early morning meal.

How about coffee?

Yeah and coffee. But we never made an awful lot of coffee, Jimmy, at that time. My father and mother of course—they used tea. But they used to have Postum for the kids. Drink cold Postum. I don't know if they have that anymore or not but they used to have that a lot and they used to put chicory in their coffee. It would come in a stick.

How did you get bread?

Well, we bought our bread along. A lot of the lock tenders' wives baked bread and there was two or three that had stores along the lock tenders.

How about pies, did they bake pies?

And the same way with pies. Along the canal we would get those. And we would get them for 10 to 15 cents a pie and they were a nice-sized pie.

Did you eat chicken along the canal?

No, we never had chicken along the canal, Jim.

Some of the canalers told me that along the canal they would catch a chicken. It would get to near the towpath and they would scoop it up whether it was theirs or not.

Well, they probably did because a lot of them did. I know the different farmers my dad knew along the canal—they never cared if we went and took the corn because we would leave a bag of coal for the corn or potatoes, whatever we needed. And a lot of them had them sitting in a basket so we could

The outside rows were for the boatmen. Corn, watermelons, and fruit frequently found their way aboard boat. Location: near Stewartsville.

take what we wanted and leave a bag of coal behind so that we were never accused of stealing. Just trading.

Just swap it?

Swapping, yeah, and if we stopped anywhere and had a boat tied up a lot of times we'd know people and they were friends of ours—why we would stop at their place overnight. We used to love that because a lot of times then we'd sleep in a house.

A lot of people see these canal pictures and of course, they are all taken in the sunshine under ideal conditions. Can you tell me what it was like to walk the towpath when it's raining and lightning and thunder and you're cold and wet and think you'll never get to the end of the day?

You bet your sweet life! And I was scared to death of lightning and thunder storms; there was always a lot of snakes that ran across the towpath. The mules were afraid of them and would jump back or foreward, and step on our feet. They never would do that if things were normal, but on lightning and thunder and rain storms it was terrible. We was always glad to get to a place where we could stop.

You always had to keep moving anyhow? You didn't tie up when it rained?

No.

You still had to buck the storms. Did you have a rain coat of sorts, in the fall or spring?

We never had rain coats and no umbrella.

You just got wet?

My dad would throw a heavy coat he'd have on the boat out to us and then we'd put it on. We never had rubber boots or anything, but our shoes were always those real heavy shoes that really could take the weather. When they got wet they were real hard and stiff the next morning from being dried under the stove.

Did you ever get frozen in the ice in the early winter?

I don't remember us ever doing that, Jim.

But some canalers did?

Yes, they did.

Did you ever have a mule die on you?

Oh yes, we had one die at Rockport.

What would the procedure be?

Then, well, it got sick on us and we had to tie up there overnight because it laid down on us on the towpath. Then we tied up right there where it went down. It died during the night and the next day we had to go right on to Washington and get Mr. Frome to go get it and then we borrowed a mule and went right on to Phillipsburg.

Then you went right on to Washington with one mule though.

One mule.

That's a little distance. You couldn't go too far?

No, you wouldn't dare to.

It would be too hard on the mule and besides they didn't want you to do it.

They didn't want you to anyhow.

And then you borrowed a mule. Who did you borrow it from?

We borrowed off Mr. Jim Camel, a colored man.

And then that borrowed mule took you into Phillipsburg and you brought it back?

We got another mule and brought that one back.

How much did the mule cost, do you remember?
Oh, about $50 or $75.

Oh, that was quite a loss then to loose a mule.
It would be a trip.

That was about what you made on a trip?
Yes, depending on how far you went, one trip.

What did your father do when he wasn't canaling in the winter, when the canal was froze and the boats didn't operate?
Well, he worked at different places in Phillipsburg. He worked for Warren Soapstone Mill here in Phillipsburg one year. Then he worked at Tippets and Woods. That was right down by the boatyard and if he could get a job in the boatyard, well, then he took the job in the boatyard.

Did he ever do anything else in the winter?
Oh yes, you mean saw and cut ice? Oh my Gosh! Yes!

Where did he do that?
Oh, he did that right above #10 Plane and he did it all down there on South Main Street, all along the front of the old Andover Furnace.

Did they ever cut ice on the Delaware?
Yes, they did, up by the Easton Bridge.

That would be north of Phillipsburg a little ways. How did they get the ice down the river then, float it down?
They didn't. They used to have wagons to bring it down. They would have mules hooked up to the wagons. They'd rent the mules. They hired the mules out and pulled it that way with sleighs or whichever they could do it with, wagons or sleighs. If it was too hard for the wagon, they'd use the sleigh to pull the ice and take it to the icehouses.

You didn't live in the canal boat in the wintertime?
Oh no. We lived in our own house then.

Did some people live on the boat all winter?
Some people lived on the boat all winter.

Ice cut from the canal in winter was stored in buildings along the berm. Location: near Washington.

That must have been kind of a damp winter.

Oh, yes.

But they had heat, I suppose, all the time.

Just a stove in the cabin.

Now Mrs. Mann, I wonder if you would sing that song you told me about before.

Yes, it goes:

Going down to Cooper's, just six o'clock
Who did I see, but ole Aaron on the dock
And he said me jolly driver, Now whose team is that?
Sure tis ole Mike Cavanaugh's, just gettin' fat.

Fal-ra-di-aiedo, far-al-de-ya
Fal-ra-di-aiedo, hum-de-dally-dae

The driver was lame, the shaft mule was blind
The lead mule had a cobcorn sticking out behind

Fal-ra-di-aiedo, far-al-de-ya
Fal-ra-di-aiedo, hum-de-dally-dae

Pete was at the hinges, Patsy at the bow,
Mike was at the tiller handle, showing him how

Fal-ra-di-aiedo, far-al-de-ya
Fal-ra-di-aiedo, hum-de-dally-dae

Going down the "17er", we were doing mighty well,
When the mule broke the towline, the boat went to Hell.

Fal-ra-di-aiedo, far-al-de-ya
Fal-ra-di-aiedo, hum-de-dally-dae

The Andover Furnace near Phillipsburg was built in 1856. Picture ca. 1870.

Now, going down to Cooper's; What did "Cooper's" mean?
Cooper's Furnace, and at that time it was called Cooper's Furnace. Later on it was called Andover Furnace.

Who did I see but old Aaron on the dock; Now who was Aaron?
Aaron Vough. He was supervisor or boss on the canal boat.

Whose team is that "Sure tis old Mike Cavanaugh's"; I assume it was Mike Cavanaugh's boat?
Yes.

The driver was lame; now who would the driver be?
That was Pap Smith. He only had one leg.

And the lead mule; Was there one mule that went ahead of the others?
Yes.

And the shaft mule was generally the one that was in the back?
Yes, in the back.

And he was blind.
Yes.

And Pete was at the hinges; who was Pete?
Pete was my father. He worked for Mr. Cavanaugh at that time. Peter Lenstrohm.

Patsy at the bow?
Pap Smith.

And Mike was at the tiller handle showing them how; Who was Mike?
Mike was Mr. Cavanaugh, the man that the boat belonged to.

Going down the "17er"; What does the 17er refer to?
It's a long level. It was the 17 mile level on the canal.

The mules broke the towline and the boat went to Hell; Why did the boat go to Hell in this case?
Well, the towline must have broke up on the boat or close to the boat near the front and the boat went into the currents or into the banks.

RECIPES

These recipes were favorites of the Peter Lenstrohm family who boated on the Morris Canal from the middle 1880's to 1915:

DOUGH DAB

4 cups flour
teaspoon of salt
½ cup lard or crisco
5 teaspoons baking powder

Stir all ingredients while dry with spoon. Put in enough milk to make a stiff dough, or a good biscuit dough, roll out on board or table and cut in square or round pieces.

Put in a greased frying pan, and fry until done. Turn back pieces as you would pan cakes. They will be brown on each side.

Note: When it was too hot to bake bread in the cabin, dough dab was often made in a frying pan on top of the boat near the hinges. Dough dab was often used as a bread substitute.

PORK FLOAT

Take salt pork, cut up in small pieces and fry in pan to get the grease out. Be sure to brown it and set it aside. Then fry 2 large onions in pork fat. Then take 6 large potatoes and dice them. 1 can of tomatoes, (cut up) and 1 can crushed corn. Cook these altogether until potatoes are done, put in your browned salt pork and season to taste with salt and pepper, bring to boil. Now make a dough as you would for biscuits and spoon in stew for dumplings. Keep covered and boil slowly for about 10 minutes until dumplings are done.

Note: This was a typical noon or evening meal for the Lenstrohm family when they were on the canal.

LARD GRAVY

Melt ½ cup of lard in frying pan, add 2 tablespoons of flour and stir until brown, add a glass of cold water or more if necessary until gravy is thick enough. Add pepper and salt to taste. (bacon drippings can be used in place of lard).

Note: Gravy whether on potatoes or on bread was a favorite meal for the boatmen and their families. It was both economical, and simple to prepare.

HUNKS-A-GO PUDDING

Hunks-a-go pudding was often made in the grease that was left from a beef roast. Now make a batter with the following ingredients: 1 cup flour (more if a big family), 1 teaspoon salt, 1 cup milk, and 2 eggs. Pour this batter in the hot grease, cover with lid, and cook on top of stove till done.

Note: Hunks-a-go pudding was often made on a Sunday when the boats did not operate. People had more time to cook. If a beef roast was not available, fat fried from suet could be used.

Made in U.S.A. by Procter & Gamble, Cincinnati, Ohio 45202
ONE CAKE
LARGE SIZE
SPECIAL IVORY SOAP COMMEMORATIVE WRAPPER
Just a little more than 100 years after our country declared independence—Ivory first floated. The first cake of Ivory was sold in October, 1879, the same month that Edison's first lightbulb lit up the night.
Ivory made a promise then—that it would always keep the public trust in its value and purity. And Ivory has kept that promise, and in so doing, has maintained its role as the favorite soap in America.
IVORY SOAP—A GOOD VALUE THEN,
A GOOD VALUE TODAY
99 44/100% PURE
IVORY
REG. U.S. PAT. OFF.
SOAP
PROCTER & GAMBLE PRODUCTS
EST. 1837
Procter & Gamble,
NET WEIGHT
9½ OUNCES
IT FLOATS

Chapter Two

The Morris Canal and Ivory Soap

Among the many mirth and melodies that prevailed with the boat captains of the Morris Canal, there was always a good supply of difficulties and problems that at times it seemed almost impossible to solve them.

As some of the boats were used as an all-the-year round residence by some of the smaller families, all cooking and washing was done on the deck of the boats. Water was plentiful and by the use of the deck pail, which consisted of a ten-quart wooden bucket firmly fastened to the end of a half-inch rope about ten feet long, they could scoop the water from the Canal and deliver it to the wash tubs without any trouble.

Coal was plentiful as all the boats loaded at Port Delaware were loaded with coal and the little straight cylinder one lid stove with its short stove pipe always had enough fire and heat in to do any cooking or washing that was required at any time.

However, one of the difficulties that these boatmen always had was the tremendous loss of soap. The brands of soap that the boatmen and their families were using was of a kind that when the piece they were using would slip from their hands it would fall on the deck of the boat and slide into the Canal and immediately sink to the bottom of the stream before they had any chance to recover their piece of soap, and the losses of soap never ceased.

At the head of No. 8 Plane west there was a Canal store. The proprietor of this store catered chiefly to the trade of the boatmen about the year 1891 when boating on the Morris Canal was still quite busy. The proprietor of this store decided to put in a new brand of soap, so he selected Ivory Soap because of the color. The color, being white, he thought, it could be seen clear at the bottom of the canal when a cake would fall overboard, and be fished out with a small scoop fastened on the end of a light long pole.

He persuaded some of the boatmen to try the new brand of soap, advising that on account of the color he thought the loss of soap would not be so great. So, yielding to the suggestion of the proprietor, a few of the boatmen purchased a cake of the white Ivory soap merely for a tryout.

At this particular store, as was always the case, the captains made their purchases before the boats were pulled out of the plane car. This gave them easy access to the boat deck. As one of the captains was completely out of soap, he immediately began to take the wrapper off his new brand of soap before the boat had cleared its cradle. Having reached the hinge section of the boat where all cooking and washing was done, he began using his new brand of soap. When he just about got the soap wet enough to create a good lather, it slipped from his hands and into the Canal it fell. Quickly grabbing his scoop and going to the side deck of his boat to get the soap before it reached the bottom. To his amazement the soap was still afloat. And with a sigh of relief he scooped the cake of soap from the water.

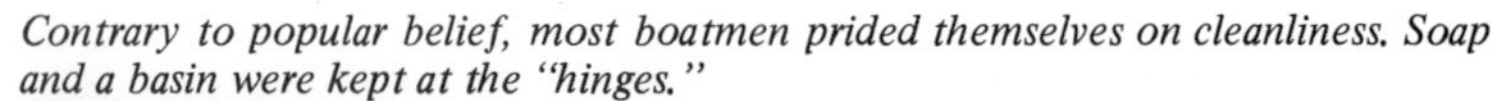
Contrary to popular belief, most boatmen prided themselves on cleanliness. Soap and a basin were kept at the "hinges."

Perfectly satisfied with this discovery, he called to the towpath boy for full speed ahead, which was about 1½ miles an hour.

He had gone only a few hundred yards, when he met an old buddy boatman coming around the bend towards him. As the boats came side by side in opposite directions, the captain who had purchased the new brand of white soap held up the cake for his buddy captain to see, saying, "Look! I have a kind of soap that floats. Just bought it at the store at No. 8 Plane." The newly informed prospective customer decided that he too would try out the white soap that floated.

In both directions of that waterway the quality of the white soap was imparted to every captain and, where it could be bought. The proprietor's soap business became so great that a representative of the soap concern came to inquire why he did such a big soap business and the proprietor gently informed him that the canalers discovered "Ivory Soap, It Floats". Then and there the slogan was born.

Along with this one incident are many more that existed along this waterway, and for ninety years of navigation there were summer and winter deviations, trials and tribulations, love and heartbreaking romances, fights and frolics, that even the words of the poet do not begin to describe the beauties and charms of this waterway.

The captain's wife is hanging out her wash on a line stretched between two towing posts. Location: Plane 1 East, near Ledgewood.

RUSSELL BATSON

Age 79 in 1976. Born at Stanhope in 1879. Worked for five years as a section hand. Now living in Byram Township near Stanhope.

Chapter Three

Russell Batson

"I had a pair of sneaks on the first day I started on my own. I was going along not thinking and I nearly picked up a copper-head pilot. I did. I fixed him. I snapped his head off with a grubbing hoe."

What is your name?
My name is Russell Batson.

How old are you, Mr. Batson?
Lets' see, two score and ten, that would make it 79 years old.

79. When is your birthday?
December 30. I was born in 1897.

Oh, you will be 80 years old then December 30. You worked on the Morris Canal you told me. Do you know what years that was?
I can't recollect, but I was 13 years old. I remember that.

How long did you work?
I worked a full four years and a half.

Then you were about 18 or 19 when you quit. What kind of work did you do?
We was a section gang. If a bridge broke down or washed out or anything of that sort, we was there ready to remend it again.

And how many were in your gang?
Five.

Do you know their names?
Bob Wolverton, Harry Hildebrand, Share, and Share, Jr. If they wanted any men that ran the planes, if it was important, they certainly would call them to help us.

Who was the boss of your section gang?
Jim Powers.

Where did all these men live? In the Stanhope area or Hackettstown area?

In the Stanhope area, the Waterloo area, the Hopatcong area, Shippen Port, and Ledgewood.

Did you have a number for your section or was there some designation?

No. They used to call it Fluke's Lock. That was the fourth lock coming east from Hackettstown.

From there to how far would you work?

Dover.

All the way to Dover. You took care of that section between there and Dover. Did you work the year around?

Only in the summertime. No boating in the wintertime.

But they didn't maintain or repair anything then?

No, no, not a thing.

When you were finished in the fall then, did you get any money from the Company or anything?

No, we just got paid off for what we done and that was it.

And there was no Social Security or Unemployment Compensation at that time?

No, no Social Security at that time.

Then how did you live through the winter? Did you have other jobs?

You had to take other jobs. I remember one time I just ate mush and milk. We used to call that chicken feed and milk. We had fried mush for breakfast, soft mush for dinner, then we'd turn around and have fried mush for supper.

Boy! Those were pretty tough times. What kind of work would you get in the wintertime when you were off the canal?

Filling icehouses. Waterloo icehouses. Then we would go from there to Hopatcong icehouses, (that's at Lake Hopatcong) up to the Pocono Mountains filling icehouses. They charged us 25¢ for a meal and board and after that, after all the icehouses were filled, if you ever see Wild West City in Stanhope, in the barroom, that was it. You'd be there for about an hour or so and out comes one fellow out through the swinging doors right into the middle of the road.

Breaks on the canal were not common, but when they occurred, they hindered navigation until repaired.

Well, that's interesting. Getting back to the canal though, what kind of work was on the section? Did that take care of everything from building bridges to patching up leaks in the towpath?

Patching up leaks and also you had to mend wood on the front end of the plane car and you had rope that was about 2" in circumference and if that rope broke, you would have to splice it with a marlin spike. And sometimes it would take 3 or 4 men all day long to splice one rope.

How often did they change the ropes on these planes?

They never changed them. No, and the plane cars go down in the water and maybe those ropes would be there all winter long. But they would pull the plane car out of the water.

Who was the Superintendent on the canal when you worked there?

William I. Powers.

Was he any relation to Jim Powers, your boss?

Brother.

He was a brother. Do you have any personal experience on the canal that stands out in your memory more than any others?

Yes, I was telling you about that. Jim Powers says "Russ, can you drive the mules single?" I said, "Oh, I'll try." I had one mule that was pretty nasty. She could kick with her front feet as good as she could with her hind feet. So I said, "Well, I'll get Jen." Jen was a tame mule. I got one canal boat ahead about a mile, so I got ahold of Jen and throwed the towline over on the other canal boat, the second one. And the first thing I knowed she says, "Heee Hawww". Bang! went the towline, and way up to her mate about a mile she went. I went up and got her and brought her back again. We fixed the towline and started again and about one-quarter mile she broke it again. And now I got nasty right there and I turned around and I pulled that canalboat from Hackettstown east to Fluke's Lock! And I told Jim Powers to put another mule in my place because I'm quitting!

And he did it?

No.

He didn't do it. You're only kidding, of course, but at least it showed him how hard you were working. Do you remember any songs on the canal, Mr. Batson?

There was only one I can remember.

What one was that?

"Oh, you rusty old canaler,
You think you're mighty nice,
Standing by the tiller blade,
Picking off the lice. Psssssst.

Would one canaler sing that to another in jest?

In jest–if one canaler had a pair of mules where he could "round" them to get ahead of him. That is the song they used to sing.

And that is the only song that you can remember?

That's the only one, yes.

Did you ever talk to any boatmen or weren't there too many boatmen left when you were around the canal?

There weren't too many left. There must have been about 5 or 6.

Do you remember the names of any?

No, I couldn't remember their names. I did know one, but I forget what it is now. But he had a pair of mules you could turn around and the mules were so thin you could put your coat on its hipbone.

Is that right. How old were you when you got married?

When I got married I was around 24.

You told me you married a canaler's daughter?

Yes, a canaler's daughter. It was at Fluke's Lock. That was the old house we lived in, which is caved in now.

What was your wife's name before she was married?

Josephine Fluke.

Her father was the locktender there then?

No, her father was the boss on the scow.

Oh, then who tended the locks?

A sister by the name of Louise, Josephine, my wife, and Maude would take care of it and Maw Fluke and there was another one, Rusty Fluke. That was it, four of them. And the mother took care of the locks.

Four daughters and the mother took care of the locks?

Yes, yes, yes.

Then that was an all . . .

Family affair. There wasn't anything to it. Just open the wickets, that's all. You would hear them coming about a mile with the old horn, you know, the old seashell they would blow, "Yea Lock, open the Gates". That would take him about two hours before he got there.

But they were ready for him when he came.

Yes.

I didn't realize that was an all female lock there. I thought they might just operate it when the men folks weren't around.

No, no. They worked that lock.

Mrs. Fluke and her four daughters worked that lock. Was there a Jenny Fluke, too?

Jenny Fluke, she lived in Hackettstown.

The lock tender's daughter uses a pole to push against the drop gate to assist its fall. Location: Flukes Lock, near Stanhope.

Yes, that name has cropped up. That's interesting.

That was the Stanhope Plane. But Jenny Fluke never took care of the plane. Now this here Willard Fluke was my father-in-law. And Charles Fluke was his brother. He was the one who took care of the Stanhope Plane.

The Flukes then all worked on the canal?

Yes, for years. He was a carpenter for the whole section, Charlie was.

In your years of occupation working on the section, did anyone ever play any tricks on you? Or did you ever play any tricks on anyone else?

Oh yes, I used to play tricks every once in awhile. I remember one time we had a washout down at Ledgewood and we was loading ashes just in Stanhope by a house that had plenty there and Dan McConnell had hard cider and the other 4 began to nip on the hard cider. So finally I said to Dan, "How's chances for another demi-john full of cider?" He said, "Help yourself." So I hid the other jug in the drum that I filled up. I put it down in the cabin. Now up in back of Port Morris roundhouse going up, I let the mules go and I saw a gartersnake right in front of the mules. I got the gartersnake as the plane car was going up. I said to my father-in-law, "I'm going to cork this snake up in the demi-john which I did. And up in back of the roundhouse Bob Wolverton, he went down in the cabin. He was half-drunk then. He went down in the cabin, got this jug and hollered to me, "Here, Booner, here's looking at you." He flipped the jug up over his shoulder and the snake came out head first! He went in the canal and I had to go in and get him or he would have drowned.

Yes, he must have gotten the shock of his life when that snake came out eyeball to eyeball with him.

Yes, and then another thing, this was at Fluke's Lock one afternoon just as we were coming home, one mule, he Heee-hawwwed and went in the canal and took the other mule with him and my father-in-law (course, I was a little devil then—didn't make no difference), he says, "Here's a knife. Go down and cut the traces." So I dove down and cut the traces and nearly got kicked in the head with one mule. And the other one laid there about three months. And before that, my father-in-law used to have an eel rack down by the house and a shower come up and after the shower, we put the eel rack in and we caught eels. Hey! 50 to 100 at a time and we put them in a hogshead barrel. So I used to eat eels then. So when Jim Powers says, "Bill," (Bill's my father-in-law) " 'bout time to get that mule out of the canal." He said, "Yes." So we went down with a block and fall and tackle. I dove down and put a half hitch around the mule and pulled him out of the canal. I never saw so many eels in my life as I did on that mule! I said, "No more eels for me." I quit eating them right there.

Mules, when they fall in the canal, are pretty helpless, aren't they? If you don't get them out right away, they are liable to drown.

Yes, because of the harness on to them.

But they say a horse doesn't lose his head like a mule. In fact, on a hot day, a horse will jump in to cool off.

That's right.

But a mule doesn't care much for water.

And every time this nasty mule—every time we'd get a load of dirt, that "Fan" would balk. Now we had the awfullest time one time, my father-in-law (I don't know what he done), but be come back finally and all of a sudden I was on top of a load of dirt on the wagon and my father-in-law says, "Hold fast now, Russ." So I held fast and first thing I knowed that "Fan" kicked the whole front end apart and I was left there with one mule and a load of dirt. She run down the towpath three miles from where we was, from the feeder lock at Hopatcong down to the mule shed. That was something!

And yet most of the times when I interviewed people they would wind up saying, "those poor old mules" or "those good old mules". There was a certain amount of affection for them.

Yes. At the same time that one "Jen" here. Now "Fan" was he a nasty mule. When I would go to feed her or clean her stall out, if she was in her stall, I would have to use an icepick to keep her away. If she got at the end of that icepick she would know who it was.

Fly harness made life more comfortable for the mules. A close look also reveals a feed basket on the rear mule. Location: Plane 2 East, near Ledgewood.

How often did you feed your mules?
When it was feeding time.

When was that? Do you remember?
When you ate, 12 o'clock. We'd feed them in the morning at half past eight. We had a fellow at the barn that would take care of the mules.

What time would that be?
That would be about 5 o'clock in the morning. And then we would be ready to go at 7. And then we would take the feed with us and the nosebag and put them on. My father-in-law never told me when I was first drove them mules, he said to me at 12 o'clock, "Russ, go down and put the nosebag on those mules." So I went down and the first thing I knowed I started to take the bridle off this here nasty mule, she hauled off and she knocked me one with her head and I went down kerplunk. I said to myself, "I'll fix you." I picked up a whiffle tree and then I hit her right on the end of the nose. If you ever seen me drop a mule, well, I dropped that one.

You took along a cut mess then probably; cut oats and feed that was all mixed for you?
Yeah, yeah.
Would you feed them in the afternoon?
At noon and at 5:30 while we quit and went back up to the barn and keep them there to feed them.

Did the S.P.C.A. ever check your mules that you know of?
There was no S.P.C.A. at that time.

There were in the big towns, but not in the smaller ones. I remember in the big towns that some of the canal men told me that every now and then they would come down and check their mules, like going through Newark or somewhere.
Not through here anyhow.

But for the most part there was no mistreatment of the animals because, after all, that was their mode of power and they had to have them. Do you remember any tricks that anybody played on you?
No, I was too foxy for them.

I know, but no matter how foxy you are, sometimes somebody else is a little foxier than you.

I was just foxy enough that I watched them all the time.

You were always on your toes. How did you work? Did you ever contract your work out, you know, like when I'm done, I'm done? Or did you have certain hours? Generally, you had hours, I suppose.

I was telling you about the time we was down in Waterloo, about William I. Powers. He happened to show up at 10 minutes past 12.

That was the Superintendent?

Yes, he was Superintendent. And Jim Powers, his brother, said to us boys, "Now just as soon as you get this overflow fixed, you can go home." It was about half past twelve and his brother came along, the big boss, and Jim says, "Now I got another job for you boys," and I said "Mr. Powers, you told us if we got this job completed that we can go home." William I. said, "Jim, did you tell them that?" "Yes." Jim said. "Then let them go home," William I. said. He had another job for us.

He was a man of some principle then.

Yes, he was.

How many hours did you work then when you worked a full day?

A full day? About 8 hours.

Only 8 hours at that time?

Yes, yes.

You know a lot of jobs worked longer.

Oh yeah. Eight hours a day. Sometimes if we had a job called out maybe in the middle of the night, you know, you had to go; take care of washouts and like that.

How much a week did you get?

Six dollars and some odd cents.

That would be for a week, that much?

It wasn't too much.

That's pretty good though for those times. What did you have

to do with your money then when you were between 13 and 18? What happened to your money then?

My father would say, "How much did you make today?" and I told him. He would say, "Let's see, you're going with a girl now" (That was Fluke's daughter). "Let's see, here's 50¢." I said, "Yeah". He said, "How much is the movies?" I said "15¢, so he said, "Twice 15 is 30–don't forget to bring back the change in the morning." Come Sunday morning if I didn't bring back the change, he'd knock me clean across the kitchen.

Well, that was common practice then that the children under 21 had to turn their money over to the parents if they so requested.

That's right.

And whatever he gave in allowance to you, that was it.

Yep. And then during planting seasons, going to school I was, he said, "Russ, I want so much of the garden spaded up tonight and if you don't stay in that school, I'll spade you tomorrow morning."

Oh, I'm amazed you got so much money at that time. That seems like quite a lot.

It wasn't too much. And you worked hard.

Yes, I know you worked hard. Did you work Saturdays and Sundays? Or only for emergencies?

Saturdays. We worked Saturdays a half a day. Sundays, no.

And now when you had a washout, what would be your procedure? Where the water is coming out, what would be the first thing you would do in a washout?

We would go up and see how much dirt and stones we wanted. That would be the boss' job.

Was there stopgates you had to close?

No. If you had a washout, say, from Netcong to Port Morris, that whole lake would wash out.

There was no stopgates in there?

No, no stopgates.

I see. If there were stopgates, you would have to close them?

You would have to close them. Where it washed out from the canal, what's to stop it?

Yeah. Even though they didn't have locks, at some places they had stopgates where they would just close their gates and then they could work on that section and not have the water waste away. But they didn't have any in your section?

No. Up here in Netcong they have a dam now, an overflow. It'd go down just so far and that's it.

Yes. And then the boss would see the break and he would have to order the material, wagonloads of stuff and then you would keep pounding it in until you stopped the leak?

Yes, knock it down until it was okay to fill up again.

Did you ever have any harrowing experiences on the canal? Did you ever almost drown in it or step on a snake or anything a little out of the ordinary?

Not out of the ordinary. These other five men were mowing on the berm side of the bank from Waterloo nearly into Hackettstown. That was cutting the old brush down and throwing the old brush out further and cutting the new on down. They wouldn't go down to the bottom of the bank, but I would have to go down. I had a pair of sneaks on the first day I started on my own. I was going along not thinking and I nearly picked right up a copperhead pilot. I did. I fixed him. I snapped his head right off with a grubbing hoe. And they wouldn't go down the bank. And then when we run into the hornets' nest while mowing. The first thing you know we would go along mowing right next to the water and we cut right through the hornets' nest and that's when "Russ" went into the canal! Clothes and all, and they'd be sticking to you when you came out. And Harry Hildebrand couldn't swim and when he hit the yellow jackets' nest, all you could see was arms and legs aflying!

I'll bet. That was a tough thing when you were cutting brush to run into them, I suppose.

Oh yeah.

How about stories? Were there any stories of people, any ghost stories or anything or legends that were handed down to you from somebody else?

There was a legend from Fluke's Lock up to Stanhope Lock. My mother-in-law used to say about half way up she always used to hear (an old man lived in the woods here); she used to hear a bell ring at nighttime. And my wife and I and her sat in there half the night to hear that bell ring, but I never heard no ringing. I think my mother-in-law had a little "bees in her bonnet".

Well, in her imagination maybe she did hear it, you know. What kind of bell would it be? A bell like that on the lead mule of the team or another bell?

A cowbell. She thought she used to hear that many many times.

Probably her imagination, but nevertheless, it was real to her.

Another thing, now today you say like automobiles driving. Walk! We used to walk from Fluke's Lock down to Waterloo and come out at my brother-in-law's in that big house and we would have a square dance. Now we walked about five miles to go square dancing and here if you go a half a mile today, you don't walk, you drive.

Right. Well, everything has changed. Just like when you were a young lad you gave your father money and now today the fathers all give their children money. That's why things are out of hand today.

They always say, "Time brings forth many changes."

That it does; not always for the best though.

Oh, no. That's right.

Did you ever see anybody drown in the canal?

No, but my mother-in-law had one boy that drowned in the lock near the lockhouse. He was the lone son. He fell in the lock and drowned. Where the house is now, you know the half of the side there? There was a fellow that lived there by the name of Harry Noland. Now I know that house from "A to Z" inside and I stopped there one day and he was canning tomatoes–cold packing them. He must have had about 120 jars cold packed then. There used to be a big garden out in the back. I belonged to the Fire Department then in Stanhope and they said, "There's a fire out in Fluke's Lock." So we went down and Harry burnt up, see. Another thing is you should see them come down there for cold packed tomatoes. And there was an image of a man where he went right through the partition and his arm stuck out of the window. That finished him there. But that partition was thin anyhow, see, and there's where he was. That fixed him.

Did your mother-in-law and her experiences there at the lock with her daughters–did she ever say that any of the canal men or boatmen gave her a hard time?

No, no, they were all gentlemen.

William Baxter, plane tender at Plane 4 West at Waterloo, with his wife and daughter, Florence. The dog's name was Uno.

They were all good?

Yeah, yeah, they got along fine. Yeah. And her father, my mother-in-law's father, took care of the Waterloo Plane for years. Old Mr. Baxter. And I'll tell you another one you'll want to reminisce with lives over here in Netcong. She is about 90 years old and relation of hers worked on the canal, Jim Kerr and Harry Share, and she could tell you an awful lot. That was before my time.

What's her name?

Ali Hill.

Ali Hill. Do you know where she lives?

Yeah.

Her address, I mean.

No, but I can show you where she lives. She would be tickled to death.

Would she?

Yes, she would. 90 years old and she's got a good memory.

You say your employment with the canal company generally terminated at the end of the boating season. But sometimes there would be an early winter and ice would form on the canal. Did you ever have any canal boats stuck on the ice in the canal?

No, but I can remember it was told to me that they would take the scow and put stones up in the bow of the boat. They then had a plate which would be fixed around and they would call that the ice-breaker. Where it would break the ice ahead of the canal boat till the next section. Then at the next section, let the other fellows take care of it the same as we did. And they called the old scow the ice-breaker. That was told to me years ago.

You just had to get it off your section and then the other fellow–he had to get it off his section.

That's right.

If he could?

If he could and then they would, no doubt, maybe the boss would go ahead and let them know that the canal boat was coming through so they could get prepared for it and keep going.

So that is what they did then in the wintertime. I have been making a collection of toys that children might have used on the canal. Do you know of any simple toys that maybe I don't have a record of that canal men may have given their children to play with or made for them or anything?

No. I remember that they used to play with them, but they are all gone now. You know what I mean. You never think, you know, like if you think today of what happened in the past, you would have kept them.

Sure.

But, no, no.

Well, is there anything else now that comes to your mind that we haven't talked about that might be interesting to people listening to this recording?

Well, I know we used to have a bad time up in Port Morris because they had a board fence between the railroad and the towpath. And there is where we had an awful job if the drill engine was working back and forth. You had an awful job getting the mules past that board fence 'cause some of them would turn around and go back the other way and you would be in an awful mess. Towline under the bow of the boat and wound up maybe in the back end where the tiller blade was. And then sometimes the mules would get away from you.

Why didn't they like that board fence? What was the exception?

That was a partition between the railroad and the canal. It was only about ten feet away. A partition where the mules would go along and if they were drilling cars out there in the yard, back and forth, back and forth, I tell you, they would turn around and away they would go.

Was it the noise?

The noise and sometimes the engineer would blow the whistle just on purpose.

And that is what they didn't like?

No, no.

They would skitter then or they would be liable to kick or anything?

Oh, anything just to get loose and get out of there.

Yeah.

And I know one time there by the roundhouse there used to be a catwalk. And two boys were in swimming and the third boy went in and he couldn't swim. And here it was not much less than six months ago that this fellow comes to me and says, "Russ, do you remember when you got me out of the canal when I nearly drowned?" I forgot about it. And there were two other brothers with him. They were twins. He told his brothers about me getting him out of the canal while we were at the Veterans Post. I'll tell you–I got potted!

They took care of you. They gave you a reward. Years later you got your reward for taking him out. Well, you certainly seemed to have enjoyed your experiences on the canal. You probably worked hard then, but as you look back on it, it was probably all well worth it.

Yes, And after I left the canal, I went in the Navy. Escorted Woodrow Wilson across the "ponds" of the Versailles Treaty.

Oh, you were on that ship?

He was on the *George Washington* and I was on the battleship. There were destroyers and submarines and we went across.

Yeah. Well, I think you have told me a good many things and you know that I intend to use this for people to hear and listen to and you have no objection to anything you have said in here? You have no objection to me taping this, have you?

No, and the whole thing was "mule down the bank, the boats run aroun', come along boat anyhow, forget about the mule."

Yep. I just wish you could recall some more songs, but that is the only one you recall. How about singing that to me again? How did that go now?

Oh! You rusty ole canaller,
You think you're mighty nice,
Standing by the tiller blade,
Picking off the lice. Pssssst!

J. HAROLD NUNN

Chapter Four

Why I Remember the Morris Canal by J. Harold Nunn

taken from his book
The People of Hackettstown in 1956

One summer afternoon in 1902 I decided to go for a swim at the stopgates on the Morris canal. These so-called stopgates were about a half mile east of the Main street bridge. It was a place where the canal company put up heavy timbers across the canal, at the end of each season. This was done so that boats could not navigate any further and were obliged to stop until the following spring when the canal would go into operation again.

The water in the gates was about ten feet deep and at the docks, on either side, was about eight feet deep. However, the canal at this point was only about twenty feet across. It was an ideal place to swim and dive for an experienced swimmer.

There were some fifteen or twenty boys there, when I arrived, but as the afternoon wore on, they went home, one or two at a time. Eventually I found myself left with only one small boy by the name of "Red L". He could not swim but he could dive. During the afternoon I had seen him, on a number of occasions, run and dive in from one dock and cross the gates and grab hold of the dock on the other side, and climb out of the water. He had done this so often that he became quite sure of himself.

In watching him, I had decided that he was exercising too much nerve for one who could not swim. A little later when I was over on the berm side of the gates, I saw him run and dive toward where I was standing. I expected to see him come up at the dock, where I was, as he had been doing. But he did not appear. A few seconds later I saw bubbles coming to the surface of the water from a large roll of moss, which was about ten feet above the gates. I knew just what had happened. He had turned his hands under the water, to his right and this had steered him up the canal instead of straight across as he had intended. He was fast in a large roll of moss and at the bottom of the canal.

I didn't even take time to think, but took a run and dived into the canal and swam under water in his direction. I managed to get to him but unfortunately I came up face to face to him. He immediately did what most drowning persons would do–grabbed me around the neck. I was in a tough spot and there was only one thing for me to do–knock him out. I punched him as hard as I could, right on the nose. The blood flew in every direction as he loosened his grip on me. That gave me a chance to get around in back of him, so that he could not get hold of me.

The canal was a natural swimming hole. Industries were not located along most sections making the water pollution free. Location: near Bloomfield.

How I managed to get both of us out of that huge roll of moss and down to the dock I shall never know; BUT I DID. I never saw "Red L" try to dive again until after he had learned to swim.

They Saved My Life

This was in the winter time and the Morris Canal was frozen over with about ten inches of 'black rubber' ice. There was some snow on the ground and two older boys and I started down the canal to dig out frozen apples from an apple orchard which was below King's bridge. It has always been claimed that apples that have been buried under the snow have a wonderful taste.

George Parks, who was a local boatman on the canal, lived in the old Parks farmhouse on the side of Buck Hill, near the old reservoir, and not far from King's bridge. During the winter months he always kept his canal boat tied fast, along the canal near his house.

At one time we were sliding along on the ice and did not know that Parks had cut a large hole through the ice that morning, near the stern of his boat. I took a run and started to slide out over the ice. The momentum carried me right out over this hole, which just had a skim of ice over it at the time. I couldn't stop myself and DOWN I WENT. Not only through thin ice did I descend, but also I was forced back under the thick ice. That is the last I knew until about a half hour later when I came to and found myself in my sister's bath tub, which was full of hot water and mustard.

I learned afterward that the other two boys had had a hard time rescuing me. They had used a piece of fence rail in getting me out of the hole. Then they carried me all the way to my sister's house about a half mile away.

YES, if it hadn't been for Pat Morley and Charlie Finger, I surely would have been drowned. More than one person has drowned in the Morris Canal in and near Hackettstown—but I "beat the clock."

BLANCHE CREGAR

Age 90 in 1976. Born at Port Murray in 1886. Granddaughter of a carpenter who repaired boats. Now living in Port Murray.

Chapter Five

Blanche Cregar

"We used to get in our boat and ride up the canal and down the canal and I was so afraid I would fall out because I couldn't swim. I never did learn to swim."

An interview with Mrs. Blanche Cregar by Mr. James Lee.

Today is September 13, 1975 and your name is . . .
Blanche Cregar.

And how old are you, Mrs. Cregar, or shouldn't I ask?
I will soon be 90.

When is your birthday?
In October, the 31st.

What do you remember about the Morris Canal?
Well, I don't remember too much, but I do remember that my grandfather was a carpenter. He used to repair the boats down by the pond.

Oh, what was his name?
James A. Bigler.

Would he just repair them or build them new?
I don't think so. They just repaired them down there. There was a building down there where they had that, but that is gone too.

This was in Port Murray?
Yes, right down here, right down here by this pond.

Do you remember the canal as a little girl?
Oh, yes.

What do you remember especially?
We used to get in our boat and ride up the canal and down the canal and I was so afraid I would fall out because I couldn't swim. I never did learn to swim.

Did you ever iceskate on the canal?
Did I ever iceskate! That was one thing I loved to do. Lil Funk

and I used to love to skate. Poor Lil.

She is gone now. Is that the lady that just died recently?

Yes. Thursday.

How far would you skate?

Oh, we would go up above Rockport. I never skated to Hackettstown or Washington, but we skated to Port Colden and Rockport.

Would there be skating parties with boys and girls?

Yes, there were some boys, but not really parties as such. Someone to go along with you, that kind of helped you, I guess.

I thought maybe you used to build fires along as you skated.

Oh yes, we did that. We built fires to keep warm.

Did you have old clamp-on skates or were they shoe skates?

Old clamp-on skates.

Did they keep coming off every once in awhile?

Yes. Once in awhile.

I know I tried them when I was young, but my ankles never stayed up right.

I always had pretty good ankles. I was pretty good at that.

You told me that you remembered people being baptized in the canal. Would you like to tell me about that and what year it was, about?

Right here is a little church book that it is in, but I do not know just where to look for it.

What happened? You mentioned in the wintertime?

Oh, the ice must have been this thick! And they cut a place not in the pond but in the canal right off the pond and they had a space there where they took them down there.

Who cut the ice? The elders of the church, probably?

I don't know. Probably they did.

Your father furnished the transportation for these baptisms. What in?

Yes he did. It is a big old open wagon. Oh, what did they call it?

A buckwagon, a buckboard?

The winter was a popular time along the canal. Among the activities were skating, ice hockey and sledding.

A winter baptism in the Morris Canal at Port Murray in 1894. Sketch by Kenneth Knauer.

It wasn't a buckboard. A buckboard is just a. . .a. . . oh. . .

Was it a sleigh maybe?

No, it was a wagon.

And these people would be completely immersed in the canal? The Baptists believed that when you were baptized you had to be covered right over the top, right? And they did not have a heated baptismal font in the church at that time, did they?

Not at that time.

That was a real testament of their faith when they went in the wintertime.

I'm telling you it was.

What did you do after they were baptized? What did the people do then?

There was a neighbor who lived up from the pond and they used to go there and change clothes. They were baptized in white and it, the dress, was so thin that when they come up out of the water their dress would freeze right fast to them till they got up the lane to this house that was Mr. Daniel Winters' place where they used to go to dress and change. But my mother and the people that were baptized that lived up here came home on the rider. You know at the time my dad took them down.

Would they have blankets and bundle them up till they got home?

I suppose so. Yes.

Would there be any singing down there along the banks? There are some songs that the Baptists have that are favorites of their church that are not favorites of others. One of them is *"Shall We Gather at the River"*. **Do you suppose there was any singing at the time they were baptized?**

I don't remember but there could have been.

That is the picture I have in my mind. And that is the song that seems to be associated with baptisms.

That is the way it used to be, but they don't do it anymore. They have the happy organ music, but they don't have the singing like they used to. I have been in the choir so many years that I know that.

Was Dan Winters a Baptist?

Yes, he was an older man, but his family were all Baptists. They all belonged up here.

It seems rather odd that they would baptize in the winter. You would think that they would have a baptism class in the late spring. Was there a reason? Was there a certain time of year the Baptists prefer to baptize?

No, it was usually Easter time, but now they baptize them whenever.

Whenever they get enough?

Yes, I guess.

I have heard of baptisms on the canal but you are the first one that had an eye witness account of it. Was it quite a spectacle?

Yes.

That is quite a thing to see–someone going in there in November. Was this November that some happened?

Even later than that.

Would it show it in the record of that book, do you think?

Yes, I imagine.

After they were baptized, and those of them that bundled up afterwards, would they go to the church for refreshments or go directly home?

No, they would go home.

I thought they might stop for hot chocolate or tea or something.

They did not. I know my dad took my mother up. We lived up to the other house then. Oh! Times have changed.

Indeed they have changed. It must have been quite a thing in Port Murray to have a canal, an inclined plane, and a boatyard. It was a highway of sorts. The local merchants got a lot of their supplies by canal, didn't they?

Yes. Just before the plane went down a Mr. John R. Robeson

There were a large variety of bridge types over the Morris Canal. Location: Port Murray.

used to have a store and that is where they got their supplies there. They belonged different places, but that is where it was here in Port Murray.

And that is where the Antique Store is located now just before you come to the plane, or was that down further?

On down further–just over the inclined plane.

Isn't that where the old trolley stop used to be?

Yes, this side of that.

Was there much lamenting the fact that the canal expired? That was 51 years ago. Do you remember when the waters were drawn off the canal?

Has it been that long?

Yes, 1924. Now they may have not gotten to it until 1925, but actually they abandoned the canal in 1924 so you see even '25 would be 50 years ago they were underway of destroying it. Do you remember when the water was left out finally for the last time?

Some. I don't remember whether I was here or there. I was busy with children at that time–raising a family.

How many children do you have?

Three boys. They are all living today. I have grandchildren and great-grandchildren.

Do you remember any particular story about the canal that maybe somebody told you when you were a little girl?

No.

Was there something that may have scared you, fascinated you, or delighted you? Somebody drowned or any tragedy? Do you remember going on any picnics on canal boats?

I don't remember that, but I have heard tell that my father, when he was a young man, used to boat. That was before he was married.

What was your father's name?

My father's name was William. William J. Bigler.

Do you remember the plane tender at Port Murray?

I remember the last one, I guess, Mr. Sutton, Mr. Casper Sutton. I think Mayberry, William Mayberry was at one time.

Well, he worked on the maintenance of it. He could have been a plane tender, too.

I am not sure of that, but they lived down there because I went down there.

Do you remember the washing machine that Casper Sutton had that was hooked up to the canal?

No.

But he had one. There is a picture of it in the book. He harnessed a leak in the canal and ran his wife's washing machine, which was quite a thing, you know, to think he got free power like that–no electric or anything. Do you remember when the fire burnt the plane house down? That plane house was different than other plane houses on the canal. It was only a two story structure. Somewhere at the turn of the century it caught on fire. They say the sparks from an engine passing quite close by burnt it up. The plane house burnt down, but they rebuilt it again in time for the opening of the canal in the spring. I was wondering whether you remembered when the fire burnt it down.

No, I don't remember that.

I have a newspaper account of that, and just wondered if you

remembered.

No, I don't.

Well, is there anything else you would like to say about the canal?

No, not that I can remember.

Could I take a look at that book and see when the baptisms were?

This is from the beginning of our church, and I have it over here

From the records of the Port Murray Baptist Church.

Special church meeting held
February 8th 1894
Pastor Vassar Presided

Meeting opened in usual form
singing reading the Scripture & Prayer
After which the following persons
related their Christian experience
And was accepted by the church
Ella. Anna & Flora Hoffman
Emily Perry   Susie Bigler
Mrs Wm Bigler   Arvilla Bigler
Eva Gardner   Carrie Molatt
Milton Molatt   Judson Gardner
Meeting adjourned with Prayer
Wm Larison. clerk

Sunday morning Feb 18th the above
persons was Baptized

and I like to look at it once in awhile.

Well, we don't know exactly what year that was when you saw that baptism but it had to be before 1897 when you were baptized at age 13. It was probably a couple years before that.

A couple of years? No, it was longer than that.

Longer than that? How many years?

It seems to me I was around 7 or 8 years old. I don't remember.

Blanche Cregar at age nine, at which time she witnessed the baptism of her mother.

but it tells in there and the date would be on it. I know their names are all on there. What is this page?

This is 1894; this is the beginning. Do you suppose that the girls at that time had the tricks that you might have had, sewing stones in the bottom of their dresses so that they would stay down?

That I don't know. I couldn't tell you that.

Did someone tell you how to do that or did you just figure that out yourself?

I just figured that out myself.

I have looked through from 1898 and I haven't seen just where they may have gone in.

I thought maybe I had a little card or something else.

This is '95 here.

I would be 10 years old, wouldn't I?

You would be 10 years old.

I could figure that. Course, I know my mother's name is there and they're on that side. There's about 12 or 13 that were baptized down there.

There are only 3 here. That was in January. Here's a bunch over here. "Meeting opened as usual; singing, reading and the Scriptures, prayer, after which the fine persons related their experiences and were accepted to the church: Ella, Anna, and Flora Hoffman, Mrs. William Bigler." That was your mother.

Yes. That's the Hoffman that has the peaches over there. Now how many were there?

Let's see. There were 11. Now, what year was that . . . That was 1894.

I was 9 years old.

February 8, 1894. Were they baptized one day and accepted another or were they accepted the same day they were baptized?

I don't know that. I wouldn't say for sure.

That was three years before you were baptized. You were 9 years old.

You see they were baptized down there in that year. I'll bet you it will tell you what they did in that Baptistery.

Here it tells in here about a lot of repair and outside work within the church, "the Baptistery" it says. That was apparently the baptism font. This is where they are talking about alterations and everything; what year was that? It must have been the last year they did it. That would be April 24, 1894. That would be in the final spring, they must have redone that. They are going to secure the services of architects to draw up plans and the plans are to be submitted to the church for approval. So that was the last baptism in the canal no doubt–1894.

I will just put that little book in there.

Then you will have it.

Here's some work I do too, when I'm watching TV.

Oh. Figuring out the words?

It doesn't do me any good. That's one thing sure. Some of the writing reads pretty good, but some of it is hard to read.

Yes, if you're not used to reading it especially.

We've had that book a long time. Course, they've got a new one now. That one was so big and heavy.

Is this normally kept in the church over there?

I guess they have a study for the pastor now. I think maybe this is kept there. He brought it over to me anyhow.

Well, I think it was a very good afternoon for me. I hope I didn't spoil your plans too much.

No, you didn't. That's all-right.

There's nothing else you remember about this canal business?

No, I wish I could think of things.

This is really what I wanted. This is really the important thing. I wanted an eye-witness account and you are it. That really goes back a ways–81 years. You know I recorded this. There's nothing harmful to you in this. You don't mind, do you?

Oh, no.

JOHN WILSON "TEDDY" DAILEY

Born at Broadway in 1876. Died at Phillipsburg in 1966. A canalboat captain at the age of fifteen. Son of a canalboat captain.

Chapter Six

John Wilson "Teddy" Dailey and The Old Morris Canal

As told to his son Pfrommer Dailey to Mr. James Lee in 1950

Pop always enjoyed reminiscing and reliving his boyhood days on the Canal. Some of his experiences as he related them to me were as follows:

One particular time they were boating at night (which was unusual) going east on the Seven Mile Level between Phillipsburg and Washington when his dad assigned him the task of walking with the mule. Being a boy, and a tired one at that, he jumped on the mule's back and proceeded to ride. Eventually he fell asleep and did not awaken until they reached their destination, which must have been the next lock or basin. At this point, my dad's father found that Pop (Teddy) had lost his cap and he was immediately ordered to go back and find it if it took all night. Apparently this was a desolate section and my Pop said that he had never seen so many spooks and ghosts as he did that night retracing his steps along the dark and lonesome towpath, although he did eventually find his cap.

Another incident happened at the Eastern Terminus of the Canal near the Hudson River when Pop had a boatload of coal and the boat sprung a leak. He said the old boat just settled to the bottom of the Canal with her decks barely awash and he had to get someone to come out with a boat to take him off.

Frequently my dad's brothers would boat with him and on this particular occasion they were tied up in one of the basins when his youngest brother, Smith Dailey, who was along this trip, fell overboard into the Canal. Pop, hearing the splash immediately jumped in after him and after helping him to the shore found out that he was quite an able swimmer. In relating this incident Pop always told about the people who were nearby who said they heard two slaps like someone clapping their hands, which indicated that no sooner had Uncle Smith hit the water than Pop was right in after him.

Usually they boated all day and tied up at night at the nearest basin or lock, but on this one particular occasion they had boated far into the night. The next morning when Pop went to check on the mule, he was frothing at the mouth and took after Pop—who chose the only escape available—he dove into the canal. At this point the old mule just curled up on the bank and died. Pop always said that the long trip was just too much for the old mule.

Pop never tired of telling us that he "was Captain of a canal boat when I was nine years old", which isn't too improbable considering the fact that he only went to school for a few months in the wintertime when the canal was frozen over. I think all told that his formal schooling amounted to about three or four years. He frequently mentioned Guinea Hollow Dam (now Saxton Falls) and how one time in foggy weather he "was out of sight of land for three days". At one time he was employed as a cook on the mud digger working around the lock at Green's Bridge. He used to boil potatoes, then hand them out the window to the crew on the end of a shovel. He was quite fond of his soups and stews and often said that he could make an old hambone last for a week by just adding more water to the pot.

Pop was born in Broadway, N.J. in 1876 and was the son of James Dailey, a boat captain. As a youngster he lived in New Village. He made his first trip on a canal boat when he was about five years old, but he didn't begin work as a boat boy till he was about seven or eight years old, and he didn't become a captain till he was fifteen.

Pop's father hauled coal from Mauch Chunk to Jersey City and my father went with him one time and a stevedore wasn't going to let them berth the boat. Pop said he got very excited,

The James Dailey family lived at Mountain View. Large families were common among the boatmen.

James Dailey on the work scow. The scythes were used to keep the brush on the banks low. Location: near Pompton Feeder.

but his father went over and talked to the stevedore and they were able to berth their boat. He told Pop, "Don't get into any arguments. You can catch more flies with molasses than you can with vinegar."

Pop didn't go to school too much. He went after the holidays and quit in the middle of March; even then I guess he played hooky quite often. He used to skip school and catch muskrats and skunks. He got $1.25 for a black skunk and $.75 for a striped one.

One teacher used to lick him every day whether he needed it or not. She used a wild cherry stick. She never told him, he said, why she licked him. But after school he told me he would get on an old mule and try to splash mud on her when she was walking the road to her home. Boy, he said, she gave him some awful looks.

He quit this one-room schoolhouse when he was 14 years old and never went to school again. It was there that he acquired the nickname of "Teddy".

He told me that lots of bums lived along the canal. One of them was named Teddy, the Bum. Once he said he asked a man for some tobacco and he threw some into the boat and somebody said, "You are worse than Teddy, the Bum." But in Newark, N. J. he was called "Spud".

He told me the bums lived along the canal in various places and when a boat would come along they would ask for a loaf of bread or something to eat. They did pretty good for not working.

There was another time he said when a section of canal bank gave away between a lock near Newark and another lock to the west. Presently 100 boats were backed up waiting for canal repairs to be made. "We were there maybe two weeks", he said.

He told me there were many stores located along the canal that provided things that the boatmen needed. He remembered one at Denville, N.J. at the lock. One time he said he went into the store for the first time and the man came out rubbing his hands and asked what he wanted. My father told him two bushels of oats, a hundred of feed, three cans of tomatoes, three cans of corn, and a ham, and he would pay him on the way

Chapter Seven

Edward Ferrand

"They was quite a decent group. We didn't have any arguments or nothing. If anything would happen at all, they wouldn't pout about it. They'd say, 'keep your mouth shut.' "

The following is a taped conversation between Mr. James Lee interviewing Mr. Ed Ferrand:

First of all, Mr. Ferrand, I'm recording this on tape. You don't have any objections to my taping you?
No.

How old are you, Mr. Ferrand?
Be 91 in July.

And how old were you when you went to work on the canal?
Twelve years old. My father used to go on sprees once in a while and I used to tend the locks for him. He had a collectors lock.

Where was that located?
Boonton.

In Boonton, New Jersey?
Yeah.

And you had to collect the tolls when he wasn't there and lock the boats through?
Yeah.

Could you tell me the procedure of locking a boat through coming from the lower level and going up to the higher level?
Come in the lower level, the gate has to be open.

What kind of gate was it there at the lower level?
Double gate, one on each side.

A miter gate?
With a hinge.

Then you opened that gate?

Opened that gate and put them as far as you could get them into the wall and the boat went in. And then you'd set the gates and the wickets so the water wouldn't go out. And then you went on to the upper gates and turned the wickets on up there; let the water in. You want to put them in easy at first or you'd knock the boat against the wall. So we had to put the wickets on easy so the water came in slow. When it got half way up, then you could put them on the whole full force. When it gets up level with the top of the lock so the boat wouldn't hit the wickets, (you had a pin in it). Take that pin out and there is a chain on the gates. That gate goes down. There was a big railroad track on the top of it on one side so the weight of that would make the gate go down in a pocket along side the gates or wickets and after that it was on a level with the upper level. And the mules pulled the boat on out of the lock up to the next lock or plane.

Was that a drop gate on the top or both miter gates?

Drop gate.

The drop gate was on the highest level. That's what I understand. That the drop gates were always located on the high level. That gate would drop down and the boat would float over top of it.

That gate had to be down cause it had to loose it and let it down in the pocket in the bottom level.

Now, by the drop gates there was a long arm sticking out—a big prop handle like. What was the purpose of that handle? Was that what you pulled the gates up with? There was a chain fastened on to it. Did that have a ratchet on it?

No. The ratchet was on the wickets. There was two of them. One set in seats that open up to pull them up. They was on slots and then there was a gear on it and then that would bind with the socket. You had to watch it. (I'll tell you a story of how my mother got knocked over with it.) And you had to put a pin in it when you get it up level so the gate was up like this standing up straight. You had to put a pin in and had to do it with only one hand. There was an awful lot of weight on there with all that water going around them

The drop gate is in the up position. The lock is ready to receive an ascending boat. Location: Lock 1 East, Ledgewood.

The shed at the drop gate protected the lock tender in bad weather. Location: Lock 1 East, Ledgewood.

Water striking an improperly closed wicket would send the handle through an arc of ninety degrees. Lock tenders had to be alert.

The long beams on the gates provided leverage to open and close the gates easily. Location: Lock 11 East, Powerville.

wickets. So you had to put that pin in with one hand and hold the crank with the other. If you didn't, it would come around and hit you in the chest. Sometimes if it hit you it would hit you under the chin and break your neck. Then before the boat went out, you had to let the wicket down or that gate wouldn't sit tight in the pocket.

You said you were going to tell me a story about how your mother almost got knocked off there.

Well, father used to go off sometimes on a weeks' spree and gave my mother a lot of trouble. He had a store on one side of the canal and the house was on this side. My mother started to take care of the boats, locks, and work the wickets. She went up to get the lower gates, shut and left the water in the upper gates. When she get them all the way up she tried to get that pin in, but it was too strong on her hands and knocked her over on her back. I thought she'd broke her ribs but she was all right. I said, "Hey, you don't go monkey with that lock no more. You go over to the store and stay there till I get the boats all up." Well, she went over there and when I got through, why, I went over to the store and tended the store while she went home.

You had a store also at the lock?

Yeah.

What did you sell in the store?

Everything from shoes, cigars, tobacco, to all kinds of cakes, pies and breads.

Was that homemade pies and breads?

Yeah.

That your mother made?

No, the baker used to come through everyday with a truck.

Now how about when you brought a boat from the upper level down. What was the procedure there?

To go down, the gates was open to go down. The lower gates you had to go down and shut them and see that the wickets was tight so the water when you put it in would fill the lock. And then you'd go up to the upper gates and put them wickets on easy to about half full. Then put them all the way up and then they'd wait till the water got level with the upper level. Then you'd take the pin out of the chain and pull that out. The weight of the railroad track would take that out and you had to put the wickets down or the gate wouldn't settle on top of the wickets.

You had to have the wickets flat with the gate?

Yeah.

Did you sometimes have to take a pole and give it a little start?

Yeah. Sometimes it would stick.

Now how many locks did you tend altogether?

I tended three in Powerville, one in Upper Boonton and one in Lower Boonton.

Do you know the numbers of them?

I think they traveled from 12-10.

12, 11 and 10.

Maybe more. I think the upper end at Boonton was 13, but I ain't sure.

But you did tend to three different locks?

Yeah.

Did you live there in the lock tender's house?

Yeah.

Were you married at that time?

House in the middle lock and it used to be a good lock. The boats went out into the River, Rockaway River.

What would be the difference between a guard lock and a lift lock? There was a difference?

Yeah.

And what was a guard lock mainly for?

A guard lock–that was down below. The middle lock was to keep the water from going down in the outer level and overflood the banks and make a cave-in.

They would be along a river when the river would get exceptionally high the water would dash down so they would guard that passage through there?

Many a night I put in when there was a heavy rain. You'd put the boards in to keep the water back; stop it up.

Now, here is a picture of the lock house where you used to live, is that right?

Yeah, that's the old lock house. Ain't there no more.

What was the location of this. Near what town?

Right in the lower end of the lock, lower gates.

And what lock was this, do you remember?

I think it was#12, Boonton locks. There's two trees, a big elm tree on one side and another big elm on the other. Then up here by the office, where they used to keep the books, and all for to write down the numbers of the captains' papers, there was three big willow trees right along the road.

Did you collect any money at those locks?

Oh yeah. Sometimes after dark a steam boat came up and would give you five bucks to let him through.

Oh, after the hours of closing?

Yeah. The hours used to be from daylight to 9 o'clock.

Lock tenders took pride in the area surrounding their home. Flowers often were cultivated. This was Edward Ferrand's home.

Boy, you never turned that down, I'll bet, did you? That was more than you made in a day, wasn't it?

Yeah. Used to get them Lehigh Valley checks. Only thirty-five dollars a month.

Thirty-five dollars a month and your house, that was your wages?

Yeah.

What were they paying when the canal closed? Did you work on the canal up until it closed?

Oh yeah.

Oh, you worked up until the end?

Yeah.

And that was what you got at the end–thirty-five dollars?

Yeah, thirty-five dollars.

How did you get your coal, winter supply of coal?

Stole it.

Traded potatoes for it maybe?

Sometimes to keep the stove going. I put the rest of it in the coal bin.

Well, that wasn't too much money, was it?

Look at what you get today.

How long did your father work in the canal before you? Do you have any idea how many years?

A good many. My grandfather was a collector up to Powerville.

What was his name?

His name was Robert Ferrand.

And your father's name?

William Henry Ferrand.

And you're Ed Ferrand.

Yeah.

Boy, your family then was certainly on the canal a good number of years, maybe a hundred years altogether. Do you remember when the water was left out of the canal and they told you the canal was finished?

Yeah.

Did they say you could live in that house then and pay rent?
No.

Did they tell you you'd have to move?
No, didn't say nothing. But I stayed there in the winter and went down to Boonton and worked in the steel works. They didn't say nothing and I lived there till we didn't get no more checks. Then I moved myself.

Nobody told you to go, you just left?
No. They tore that house down. That was the best built house I'd ever seen on the canal. Boy, that was a big house. I suppose they done it because of the three locks there.

How many rooms were in the house down stairs?
1-2-3-4-5-6 into it.

Altogether?
Yeah, and there was a cellar. Two big apartments all cemented on the bottom.

How many rooms downstairs, two rooms downstairs and four upstairs?
I think three. One big bedroom and two little ones went the whole half of the house. The big one, the other two was on the other half.

Did you have a boss that used to come around occasionally and check up on you?
Oh yes.

Who was your supervisor or boss?
Heaton.

Mr. Heaton? Well then, what did you do with the tolls you collected? Did he come and get them or did you send them to Phillipsburg?
Well, my dad took care of them.

Oh I see. But were they sent by mail?
Had a big ledger and the ledger was that square. We'd write down each load and the captain's papers.

But what did you do with the money?
Didn't collect no money.

Oh, you just wrote it down?

Yeah, and sent it to Phillipsburg to let them know how much they was going to load the boat again.

How would that go to Phillipsburg, by mail?

I used to have to go to the post office for my dad and used to have an envelope that long and about that wide and there was two postage stamps. It cost two cents in them days.

And everyday they were mailed to Phillipsburg and then they knew the boats were coming in?

Yeah.

Boats passing Lock No 1 West Monday May 15 1893

HOUR A. M.	HOUR P. M.	BOAT	CAPTAIN	BOUND EAST	BOUND WEST	CARGO
	5.40	752	Lowen	s		Coal
	6.10	682	Dagon		s	Empty
	6.15	500	Daily		s	s
	6.24	709	Dailey		s	s
	6.45	569	Lenzbrohen		s	s
	6.50	485	Schaible		s	s
	7.00	602	Mead		s	s
	7.05	721	Meyers		s	s
	7.30	685	Cook		s	s
	7.10	481	Major	s		Coal
	7.20	756	Dalrymple	s		s
	7.40	718	Sheals	s		s
	7.45	753	White		s	Empty

A page from a ledger kept at the Morris Canal Office at Stanhope (Lock 1 West) which recorded boats passing Lock 1 West between October 6, 1892, and August 23, 1893. On exhibit at the Museum of The Canal Society of New Jersey, Waterloo Village.

How many hours a day did you work in the locks?

Daylight to 9 o'clock at night.

When was daylight then—five?

Sometimes five and then it'd get darker towards winter and days would get shorter.

Did you often pass the boats through if they were good friends of yours? Or if you knew them well, did you pass them through after you closed up?

Once in a while.

Generally you didn't, though?
Yeah.

How about Sunday, did you work Sundays?
No.

Do you remember the 4th of July? Did the boats run on the 4th of July?
Yeah.

Was there much fighting on the canal or was that overplayed?
No.

There wasn't too much fighting as far as you know?
No.

There's a lot of talk about fighting, but everybody I talk to says "no, there wasn't much fighting". There was a fight once in a while.
No. They was nice captains.

That's what everyone says.
Some captains had their whole family on the boat.

Right, and that brings me to another question. How about children on the boat and their toys? Do you remember seeing any children's toys that their parents might have made for them? Did your father ever make any toys for you to play with?
No.

Well, did he ever make you a whistle out of a willow branch?
Oh, yeah.

Well, that's a toy. Did you ever play with buttons on a string and make them spin back?
Oh, yeah, buttons on a string.

Can you remember any other things he might have made for you?
I think he might have made a sled one time for me.

A sled with runners?
Yeah.

An old man told me about a little stick with notches on the top,

An inclined plane provided a logical hill for winter sledding activities. Location: Plane 11 East, Bloomfield.

Locks were a natural attraction for children because of the activity. Location: Lock 8 East, near Denville.

and a little propeller on the top and when you rub a penny over the notches that would spin around. Have you ever seen any of those?

No.

I wish I had brought one along to show you. What games would they play on the boats. Would they play jacks?

Oh, yeah, with a little ball.

Did they have homemade dolls?

Yeah, they played with them. They used to be in the middle of the boat, be quite a space. There used to be two chests there for feed. Hams, potatoes and onions and stuff they put there. Used to be two chests on each end of the boat. They used to be hinged up, ya know.

Do you have any experience that stands out most in your memory?

Oh yeah. One time I started to tend lock for Mrs. Peer. She used to live in a house in the store across from my father's lock. And she used to see me tending the lock. Her husband died and she come down and asked my mother if I could go up and tend the lock for her. I was between twelve and fifteen, I guess. She said, "You can have him if you want." Mrs. Peer says "I'll give him fifteen dollars a month and his eats and sleep." I stayed there and worked for her I guess three years. But the worst of it was, it wasn't my accident, it was a boy's accident. Boys that lived in the house up above used to come down and watch me put the boats through. It was towards night and I had to go get my supper. I heard a boat coming down. He was blowing his horn. (They'd blow their horn to let you know they was coming) I finished my supper quick and come out. I seen something was wrong right away. The boys, they had skidooed after they had opened the wickets on the upper end, and the lower end gates was open. Well, all that water going down the wickets was going down in the outer level. I hurried up and went over and put them wickets down quick and I went down and set the gates and I pulled up the wickets. Boy, all that water going down there made that boat come down like a bugger!

Made a current there?

Yeah, boy, I was afraid. I was shivering in my pants! I said, "That's going to go right down through that gate." I got the lock full but the boat was so near it came down, hit that head gate and cracked it. Boy! They was four or five days getting it fixed–getting boats through again. Wasn't my fault. It was those boys' fault.

What did the boss, Mr. Heaton, say when he saw the condition of the gate? Did he ask you what happened?

Yeah. Mrs. Peer went out and told him what happened too and he said, "Okay, we'll get it fixed." And he did but it was four or five days before they got it done. Cracked it, but didn't go all the way through the bow. The bow hit it right in the middle and cracked it, but it didn't go all the way through. It was lucky I put the wicket down or it would have broke that too.

Did you ever fall in the canal?

Oh, many a time. I was pretty near drown twice. My dad pulled me out once. He saw me down below the lock along the log. Used to have a log there to keep the boat from hitting the bridge. He seen me down there. I was learning to swim then. He put the wickets down and you can imagine a big lock 40 or 50 feet deep letting the water out of it. Boy, it looked like Coney Island. Boy, oh boy! The wave washed me up in a little hollow the other side of the bridge. He come down and I was a puffing and could hardly breathe. Dad came down and looked at me and walked away–seen I wasn't drown.

That was quite an experience. Did anyone ever play any practical jokes on you?

No. I don't know whether you remember ole Peg-leg Smith, the captain on a boat. I telled him, I let him thru the lock with a load of coal. I said, "You hug the towpath now so you can keep in the channel." There was a brook that run down back of the house and everytime it rained hard it made a big sand bar right down in the water. I said, "You keep out of there you'll go into the sand bar." He never steered for the towpath at all. He steered right straight–right into the sandbar.

Well, they had to work hard to get that boat loose. The mules was a-tuggin on the boat and pretty soon they broke the line. The mules went down the towpath and right into a farmer's corn patch. They was a having a good time in there. Pretty soon Pop had two big poles on the boat and he got ahold of one of them poles there. He had one in his hand then trying to get the boat away from the sand bar and he put that on the other side of that boat and put it as far into the towpath as he could so he could swing himself. He swung himself right over to the towpath. He swung himself that day and pretty soon I saw him coming back with the mules. I don't know how they ever gave him a boat of coal to go. He could never take care of his mule drivers. He used to swear and holler at them all the time. I bet he'd have three or four drivers before he'd get down where he was going.

Do you remember any of the other captains that came through there?

Oh, I could name a lot of them. Maybe some of them lived in Phillipsburg. There was Thorton, Warner, Drumple, Fox, Cook.

Do you remember anyone by the name of Lenstrohm?

Yeah.

That's my wife's grandfather. How about Poppy Pearson?

Oh, Pearson, I remember him. He was one of the early ones.

Yes, his family was on, but he was one of the last ones on the canal.

There used to be one by the name of Bill Black. He used to be on a wood boat.

Do you want to tell me about the wood boats? It used to just carry wood, is that right?

Yeah.

What about Billy Black? Do you remember anything about him? Where did he get his wood?

All along the canal. In the wintertime the woodchoppers in the woods would fetch the wood down and rack it along the canal. Now down below the lock sometimes he'd get a load there. I used to go down and help him load the boat. There's be two of us. He'd give us five dollars a day. I would load that boat and he'd come back maybe a couple weeks and there was another one. He most always took all that wood to Paterson. His name was Bill Black. Then he had a son named Charlie Black. They used to live, I guess, near Lake Hopacong. They used to go along the lake sometimes and get wood. We used to have racks of wood along the lake there.

Well, Billy Black apparently when he was a lad was quite a rounder. But he took to religion later on in life and when you knew him, was he a pretty nice guy?

Oh, he was a nice fellow, what I seen of him.

He took religion, but at one time he was suppose to be a drunkard. He wrote a little pamphlet called "Billy Black the converted Boatman." Have you ever read that?

No.

Well, I have one of those little pamphlets and he told how he went to Mauch Chunk, Pennsylvania and loaded coal and he said sometimes he would go on a drunk and lay around for days at a time. One day his wife said to him, "Billy, if you

Wood cargoes varied from fire wood to railroad ties to building lumber. Location: near Denville, ca. 1905.

don't straighten up, I'm going to leave you." So he said he made up his mind he was going to stop drinking and he never drank another drop. From that time on he thought life was a lot better for him. And he never drank another drop. And that's the same Billy Black you're talking about because he did carry a lot of wood later on in years. In fact, he carried more wood than coal; but at one time he carried coal.

Do you remember when the boats used to carry pig iron? There would be a stamp on it. Came from Wharton and up around the blast furnaces.

Yeah, the furnace at Wharton, Phillipsburg and Waterloo, down there. Yeah, there used to be a feed boat come down too, to supply all the stores along the canal with feed. Then there used to be a liquor boat that used to come down. The superintendent had a boat on the canal that was named after his daughter, Florence.

Do you know what its name was before it was called the Florence?

Katie Kellogg.

Regular inspections of the canal were made by company officials. The **"Florence"** *was used for this purpose.*

Yes, there are pictures in here of it.

It was painted white.

And what did that boat do? Did they inspect the canal and pay the men?

No, not pay.

How did you get your pay?

Oh, the superintendent.

Check? Did he give you cash or check?

Check, Lehigh Valley check.

Yes, in later years they probably did give a check. But at one time, they used to go around and pay the lock tenders. This was many years before. I guess, before the Lehigh Valley Railroad took over the canal.

Well, I guess the Lehigh Valley took to over for 99 years.

Well, they leased it for perpetuality. I guess 99 or 999, but it never turned out to be a very going profitable concern for them.

The railroad got too thick.

They took the canal over in 1871; the Lehigh Valley Railroad did. Do you remember singing any songs on the canal?

No, never heard them.

I'm trying to collect as many songs as I can. There were some songs that were sung on the canal and some that were apropos only for the Morris Canal. I do have some in my collection. I was wondering if you ever heard any of them?

Never heard any. You could hear them playing on instruments but you couldn't hear them singing.

You mentioned you would hear a horn blow when a boat was approaching. They were probably conch shell horns mostly?

Oh yeah. There was a shell with a hole in one end and they'd put that up and boy, they could make it ring. You could hear it a half mile away.

Did some boats have a tin horn?

Yeah.

Well, what else do you know about the canal that we haven't

covered?

Well, I don't know if Harding told you about it or not. When I was tending locks for my dad, he come home early one night and forgot to collect the paper off of Birdie and he got mad. He says, "You get up that towpath and collect it." Well, I went and made up my mind I was never going back. I was going to pull for the West. And I got up, I guess it was about Waterloo or I think I went on out to Phillipsburg and my mother got worried about me and had a brother Robert (named after my grandfather and who powered the local locks). She said, "Bob, you go up and try to catch him." Well, I guess it was two or three days. I went with, I think it was Phil Brady or Paul Brady and I got on the boat with him and I didn't walk no more, I rode. It must have been Phillipsburg or Port Warren. He found me one night and he coaxed and coaxed me to come back. I said okay and I went with him down to, I think it was Stanhope and we got on a caboose on the Lackawanna Railroad. He said, "Take me down to Boonton." "We ain't going to Boonton, we're only going to Denville." He says, "That will be far enough." So he put me on it and we both went down. The next morning was Sunday morning. We slept in a box car that night. Next morning there was another freight train going the other way through Boonton, so we asked if we could take a ride down. We told him all my experiences and the conductor on it says, "Okay, you can get a ride to Boonton."

What about that mule story you told your daughter?

Oh yeah. There comes a big snow storm. I guess there was pretty near a foot of snow. It was about the latter part of October or first of November. It froze the ice, about two or three inches of ice and that snow together, froze and then the boats couldn't get through. There was three boats stuck. The Super asked if we'd take the mules up and hitch onto the other mules and tow the boats down. He couldn't get nobody. I said, "I'll go." Boy, it was snowing hard. It snowed good and hard till I got to Dover. And then it commenced to rain. Boy, didn't I get wet. Boy, I had three mules, two gray and a big high–he was six foot when he was straightened up.

He was white, pure white; called him Dan. I got him and rode him. Boy, did he move. He pulled them other ones sometimes; he pulled them right along. I went, I don't know where it was, above Stanhope somewheres and I went in to explain to old Burd, his name was. Well, I went in there and got that far and boy, was I soaked—water ran right out of my britches. They had a big stove, belly stove, big one and it was red hot. Boy, I sat down in a chair and enjoyed myself. He called the girls and they fetched over apples and a jog of cider and boy! I enjoyed myself. I got dry and got the mules together again and went up on till I met the boats. Well, it was so thick, every section had a scow in it and two or three men on it. We had to put the ice breakers on front of the bow to break up the ice channel so the boats could go through it. Soon as they'd break that, why, the boats could go through. But they was afraid the boats cutting the ice without the breaker would cut through and flood. Well, I hitched on to them and I stayed with them till we got to Powerville and that first one went to the Powerville Paper Mill. The other two went to Boonton and that was the town below that. One went to Clark and the other one went to Sellen and then I was through with my trip.

Oh, you were a mule boy too, then?

Well, he couldn't get nobody. I said, "Sure, I'd go."

What did you get paid for that job?

Just regular wages.

You know you are only the second one to tell me about the ice breaker.

We used to have a big heavy timber that was iron plate on each side where the breaker hit the ice. Flat iron plates, it was. I guess six by six timbers and built so it would be about a foot apart. The mules would get on that with a towline and pull that scow along and it would break the ice in the channel.

Well then, the breaker wasn't formed in a "V"?

No, flat.

Did it come to a point?

Oh no. That thing weighed pretty near, oh, about pretty close to a thousand pounds. It was chained on to the scow and went right down deep in under the ice the whole width of the scow.

Then how did that break the ice? Did they pull it up through the ice?

No, they crushed it up against the plates on the breaker. As the mules would pull it, why, the ice would break.

And it was the duty of each section man to get the boats off his level when it froze up, if he could do that?

Yeah. They couldn't move till they broke the ice up.

It was too much for the mules? How many mules would pull that scow with the ice breaker?

Two. Sometimes we put four on.

If the ice was too thick, would that pull the scow under it then?

No.

How thick would the ice be that it could break?

Ice was cut on the canal along its entire route, providing many boatmen with winter employment. Sketch by Kenneth Knauer.

They'd fill up the back end of it with a lot of rocks so it's have weight.

Oh, rocks on the back of the scow to balance it?

Yes–right next to the cabin.

How thick of ice could that break?

Oh gee, sometimes four inches of ice.

But when you got five or six?

Then you're done for the winter. And now the other one I wanted to tell you about was the cable on the plane and the shackle, you know, that the cable goes into. That pulled apart, and Dad went up to help put a bridge in. The Powerville Bridge. It was a long bridge there and they was on their way back and got about half way down the plane when this shackle let the cable out and it come down and just scaled the scalp off my dad's head–right off. He didn't go to the hospital–he stood it. It took the scalp right off the top of his head.

Never had any ill effects from it? Got to be all-right though?

Oh yeah, He didn't break no bones in his skull. It was a tender spot. It's a wonder it didn't crack his skull in.

Was your father living when the canal went out of business?

No, he died in '41.

Well, he was still living when the canal went out of business.

Yeah, he was 81. His father used to tend the lock up in Powerville. He used to be a collector too. That was way back. He was there quite a while. He had 27 children by two wives. First wife's name was Jane Hogue and the next wife was Sarah Blanchett. She was born right down here on the flats, down there by the church. All the girls had by the second wife only lived to be about two years old. The biggest part of the first wife's boys were in the Civil War. One was killed in the "Battle of the Wilderness." His name was Edgar Ferrand.

Mr. Powers–do you remember this name?

William I. Powers. Oh, I remember him. Used to pick cherries in front of his house. Big white Oxharts. Used to take them in and his wife would deal them out.

Well, he became the superintendent and his middle name was Imla.

Is that what his name was?

His middle name was and he was a Civil War Veteran. He was wounded seven times in the Civil War. His daughter's name was Florence and that's who they named the pay and inspection boat after. Prior to that, it was Katie Kellogg.

I used to see her helping her mother when I'd take the cherries in.

I would like to get a picture of her. I have the boat but I don't know what Florence looked like.

We ain't got any either.

Did you ever get robbed on the canal?

No, They was quite a decent group. We didn't have any arguments or nothing. If anything would happen at all, they wouldn't pout about it. They'd say, keep your mouth shut. Oh. sometimes you'd let the boat down too far and you had a big snubbing post on the outside above the lock and used to put the rope around to keep the boat from going into the gates. And if you let the water out too fast sometimes that

The towline went from the shaft mule to a towing post when the boat was loaded. When light, it went to a cleat.

rope would catch and would break them. Boy, they used to get mad when you'd break a rope. They had to get a new one then, you know. And the old mules sometimes would break their rope. It got too old and would break. There was a couple of mules, I don't know whether it was at the Powerville Lock or Lewisburg, but they was coming around and a boat got stuck somewheres and they pulled and broke the line and both of them tumbled over in the canal. One of the mules got tangled up in the harness and rope. He drowned.

Couldn't get him out?
No, couldn't get him out in time, they both drown.

Mules didn't like water too much.
They had to be coaxed and coaxed along. They were stubborn.

Did you ever get kicked by a mule?
No. I come near it but I always kept out of their way.

You never got behind them and tickled their rear end then?
They most always used their back legs too, get ready and out comes their heels.

They would bite you too sometimes?
Oh yeah.

But most everybody I talked to said they always felt sorry for the poor ole mules. They had a hard job pulling that boat.
Yeah.

Most people treated their mules pretty good, too.
Oh, they did. Oh Peg Leg Smith, he didn't treat them very good though. He used to pound them every once in a while with a stick. When I was a boy there was an old belt factory where they made belts for these sewing machines. We'd go up there where they'd throw the leavins out and sometimes there would be a long piece of leather, round too. Nice. Make a good whip out of it. We used to get all the good ones we could find and fasten them on a stick and we used to sell them to the captains for a nickle apiece. So we used to have money for candy for a good while.

For a nickel, you could get a lot of candy.

First, I'll tell you about a lad falling in the lock when it was empty. He went up to the town of Boonton. He was a scow hand. He used to sleep on the scow. He come down, must have been drinking, I guess. He walked right off into the canal. My father was uptown quite late that night. He came home, he heard somebody hollering. He said, "Somebody in that lock." He went and got a ladder and flashed it down. He seen there was Bill Brown. He was standing there with his hands into a rock to hold himself up. That lock was, oh, must have been 40-50 foot deep, stripped of water. My dad says, "Hold fast, I'll go get a rope." He went and got a rope and lowered it down to old Brown and he got ahold of it and he pulled him on down to the gates and put the rope under the platform for the bridge and pulled him out and down by the towpath and put on the bank. Then I'll tell you about two people drown. It was a colored lad, his name was Jackson. He'd been uptown, too, in Boonton. And he come down towards the lock, went down the towpath where he lived down to big sand pit where the Lackawanna got their sand. Well, he come down and there was big steps all the way down the locks for boatmen to carry their towlines down to the towpath. He was sitting on one of them rocks, so they believe, and he had a bottle of whiskey was sittin there but he wasn't there anymore. He must have got tipsy and fell over off of the rock down into the canal. They hunted and hunted and hunted for him and couldn't find him. Pretty soon they drug the canal. There he floated, I bet a mile, his body was found at the gates where they let the water out to empty the level. They found him there. He traveled pretty near a mile before they found him.

Then there was another one—a big Polock, I guess, you'd call him. I forget his name. He was a big Polock, polish lad. He lived in a house above the canal. He used to come over there to out on the sand bar. This fellow got, I guess too far out. The edges of the bank gave way and he went down in the canal channel. They tried to hunt him and hunt him, the people in the house he lived in. They couldn't get him. They came up to the house when my dad came home that night. He went sliding off that sand bar down in that channel and the current from the brook throwed him over against the cribbing. They used to have cribbing all along the towpath to keep the dirt and rocks from caving in. There he laid; the pressure from the water kept him up against the cribbing.

What did you do in the wintertime?

Oh, in the wintertime I used to skate and sleigh ride. I used to skate from Boonton to Newark.

What did the canal men do in the winter? Did you get paid every month even when the boats weren't working?

No, no, just when the boats was running. I worked in Boonton in the wintertime.

Oh, you had another job in the wintertime. But you had the use of the house?

Yeah, I'd get house rent free. I used to work in the steel plant.

Did they hold some money out of your pay and give it to you about Christmas time?

No, never did me. The last check you'd get would be the last month you'd put the boats through. I could tell you some more stories you might now know about but I can't think of them no more.

About your father?

Yeah. In the winter time he wouldn't be tending the locks, why he'd go down and help Tom McMillan, the boss carpenter down there in Mountainville. Well my dad go down there in the winter and Tom would learn him how to carve out all the wood for the gates, bridge work, lock work, planes and shave the wood. Well he died and my dad did all the work for the canal. That was the lower end of the canal.

Each section had a carpenter shop generally. It was too much for one.

Yeah, they used to make gates, head gates and lower gates. Had a funny name, Tom Ed.

Tom Ed? Yes, like Jim Lee. You could use it either way.

Yeah.

Chapter Eight

New Village

New Village was a small settlement before the Morris Canal was chartered. Just how old this little community is, apparently no one knows. The book *History and Directory of Warren County,* published in 1928 mentions that the Union Sunday School of New Village was 117 years old. These figures indicate the Sunday School as having been organized in 1811, long before the Morris Canal was constructed.

In 1910, one-half the population of this small community, which was about 500, was of Italian extraction. They brought with them to this country many native customs, one of which was how they laundered their clothes.

Before the advent of the washing machines people would boil their soiled clothing in wash boilers, rub them on washboards, rinse them in clean water, wring them out, and hang them on the line to dry.

In many of the small towns and villages in Europe, and especially in Italy, water was not too plentiful. Often one pump or a common well would serve the entire town. In order to

Locks took their names from the lock tenders, nearby towns, or a special feature. Location: Gardner's, Lock 7 West, near New Village.

conserve water, the women, after they had boiled their wash at home, would take it to the local river or small stream. There on flat stones or on racks built by their husbands, they would rinse and beat out the soapsuds from the wash.

This way of laundering their clothes was, I understand, not thought of as an unpleasant task, but one that was looked forward to by the housewife and children alike. It gave the women a chance to meet their neighbors and gossip a little. Also it gave the young ones a chance to play with other children whom they might only see once a week.

Mrs. Gina Bellini was born in New Village, the daughter of Italian immigrants, Lavinia and Nazzareno Marketti. She well remembers when she, her sister, and her brothers went along when her mother carried the family wash about a quarter of a mile. Here it would be rinsed in the clean waters of the Morris Canal.

This practice was not limited to New Village, but was done in many locations along the canal.

Some individuals rinsed their weekly wash in the canal. At New Village wash day was a social event. Sketch by Kenneth Knauer.

Most planes were single track. Boats had to take turns ascending and descending. Location: Plane 7 East, Boonton.

Captains always would try and clear a lock or plane just prior to closing time, thus gaining extra boating time. Location: Diamond Springs, near Denville.

SIMON JOHNSTON

Born at Stewartsville in 1886. Died at Phillipsburg in June 1976. He was a canalboat captain and son of canalboat Captain David Johnston.

Chapter Nine

Simon Johnston

"You didn't have to know anything about reading and writing. All you had to know was which way to take the boat and steer the boat and get it to its destination. That was the only thing."

Mr. James Lee interviewing Mr. Simon Johnston.

How old are you, Simon?
How old am I? 89?

Mrs. Johnston: You are 88 right now. You will be 89 in February.

And your name is Simon Johnston and you know I am taping this. You don't have any objection to this, do you?
No, no. Certainly not.

Your father was on the canal, you said, before you. How long was he on the canal?
He was on the canal until pretty near his middle life. Then he went to be a detective on the railroad.

Oh, that is quite a switch from the canal to a railroad detective. When did you start on the canal; what year? Did you work for your father as a towpath boy in the beginning?
Yes.

How old were you when you became your own boatman, your own captain?
I was about 17 or 18 years old.

What were the requirements to be a captain on the canal boat? What did you have to know? You had to know your reading and writing, I suppose.
You didn't have to know anything about reading and writing. All you had to know was which way to take the boat and steer the boat and get it to its destination. That was the only thing.

They had rules and regulations on the canal. Were you given an examination before you went as to your ability to know these rules and regulations?

No.

You just kind of picked them up as you went along?

Yes. Picked them up as we went along. When I wanted a boat, I asked for a boat and that is all there was to it.

When you asked for a boat, the canal company owned the boat; then you had to supply the mules or the horses?

Yes.

How much money did you make on the canal? I know it varied with the times.

It wasn't very much. I would say about $50 a trip.

Was that in the later years or the early years?

We never got very much of a raise.

When you were a towpath boy working for your father how much did you make?

About $3 a trip.

Whatever he gave you?

Whatever he gave me.

When you got to Jersey City, did you have to help unload the boat or at any of the points in-between?

I never had to.

Could you do it if you wanted?

I sometimes could guide the line. That is when they put the bucket down in the boat. You could do that, and when they got it up to the right place, you could dump it. Sometimes they gave you that job if you wanted it.

How much would you make on a job like that?

About $1.

How long did it take to unload a boat?

All day, about 6-8 hours. Lots of times there were just two men. They would unload that boat in a day. They would chuck the coal with the shovel up in a wheelbarrow and wheeled it over to this dumping place just across the canal.

The Port Delaware chutes enabled an easy transfer of coal from railroad cars to canal boats. Several boats could load at the same time. September 2, 1911.

Did they always get full tonnage? In other words, if you left Phillipsburg with 70 tons, did you always have 70 tons when you arrived in Jersey City?

No, not always.

What would happen to some of it? It wouldn't evaporate?

We used to chuck it in the canal to make the boat lighter. We used to give it to people if they wanted it.

You would swap off for apples or potatoes maybe sometimes?

Yes, sometimes.

And sometimes the locktender maybe would want some for his winter supply. But nobody seemed to miss it, what was taken off?

No. They were foxy. They would chuck this coal away or give it away to make it a little lighter, so we could pull the boat easier, see. There was just nothing to it. And when we got down to the destination, then they would take a bucket and throw a lot of water over it and wet it to make it heavier. That was that.

Do you remember anything outstanding when you were on the canal? Any harrowing experience that you had or that you nearly lost your life? How about when you were a boy? What stood out in your mind most as a boy? Were you ever afraid? Were you ever scared traveling at night?

No. The only time I was afraid was up here one time right at New Village under the bridge up there. I saw something sticking out of the water and when I came to look at it, it was a dead man looking at me.

I'll be darned.

Boy, I was scared. I was just a young fellow. I will never forget that experience.

What happened to him? Was he killed or did he drown?

He stoned himself. He drowned himself.

Oh. Was he a young man or an elderly man?

I don't just remember. I don't think he was very old.

I imagine that would shake you up a little bit.

It did.

Did you ever have an accident where you fell in and almost drowned or could you swim at an early age?

I could swim.

Drowning was one of the tragedies that happened on this canal quite often.

No, I never had no experience of drowning . . . But I used to like the canal. I liked the idea of going from place to place and seeing different people.

Yeah. Now would you say that was a pretty good life?

I would.

How about as for a boy though? They started at a pretty tender age, didn't they?

I thought it was a pretty good life. It was a hard life when you stop to look at it now. When you stop to think of people living on a boat. It was just a little square place where you slept and ate and everything. I slept under the water there all the time practically, and kept my shoes in the dry.

And when you slept you could hear the water gurgling past, I suppose. I suppose those sounds still haunt you today.

No, there wasn't much sound of the water.

It was so slow running you didn't hear it?

You generally tied your boat up where there wasn't too much current.

Yet in the spring and in the fall when you boated late into the fall and started early in the spring, it was pretty cold and it wasn't very good to work then?

It was very cold for us boys. We didn't have shoes to wear. We went bare-footed.

That is what I mean. When it rained, you had no raincoats to put on.

No.

And yet you had to be out in that strife and storm making sure that those mules didn't goof off along the way. That is the part that I think about that most people don't see because the pictures that were taken were taken when the sun is shining and everything is just right. They don't show the storms or walking late at night after you have passed the lock or the plane in which you could continue well into the night. That's the part I mean.

One experience come to me when I was a boy: From Bloomfield to Paterson, there comes a bad place, you know. And we heard ghost stories and we was scared to death. One night I come up and I was walking with the mule. I would get ahold of the hames or something (I forget what it was called) and while I was asleep I'd walk with them. One night I fell over this here thing and Boy! Boy! I was scared! So, when I fell over this thing and saw it go down over the bank I was scared to death. I thought, "Well, they say there are ghosts." I was only a boy about 8 or 10 years old. I thought "I am going to see if that is a ghost." Then I went down over the bank after it. It was a calf!

A calf? A white calf?

And then I came on up (this was later though) and for some reason we were fast or something and we laid there. The level was low, I guess, and we could not go any more. I laid down on the towpath and boy, I woke up and the rats were on me—all over me.

The rats! How about water snakes? Were you ever bothered by them?

No, they never bothered us.

The canal officially was open from 5:30 AM to 9:30 PM. The evening haze fostered ghost stories. Sketch by Kenneth Knauer.

I mean walking at night bare-footed along the towpath. I've heard someone say he walked along and stepped on one. They're not poisonous, but it would give a 10-year old boy quite a start.

I did that one time in the canal. I seen something on a rock and I went down and picked it up and . . . I had a watersnake in my hand! When you stop to think we slept on them boats, to think families were raised on that little cabin . . . Now I was just saying this morning, we didn't have automatic washers, and didn't have bathtubs, and we didn't have no toilet in the house. Think what them poor boys had to do. When they had to go to the toilet, they had to go under a tree or in the bushes along the canal.

Yes, I suppose you had a pot in the cabin for at night.

Yeah.

Even the washing wasn't too good. Course, the canal water wasn't too bad, up this way. You could wash in that quite easy, but when you got down to Jersey City, that was a different story. You couldn't wash in that stuff.

No.

Did you ever play any practical jokes on your father or your buddies that you can remember? Maybe you tied the boat up so when the mules went to pull them, they couldn't pull the boat or something like that?

No, I never did anything like that.

You didn't have much time for practical jokes then.

No, you was going from early in the morning until late at night. The minute you stopped you fell in bed and you were asleep.

You were tired.

Didn't make any difference.

How about on a Sunday, Simon? The canal didn't operate on Sunday. What did you generally do on a Sunday if you were out of town?

If you got on a long level, like the 7 mile level or the 11 mile level, or the 16 mile level or whatever it was, you would try to get on that and tie up and give the mules a rest at night and then in the daytime, you started out and . . .

Leisurely go along till you came to the next one?

You would go along through the week. If you got there on a Friday night, you would go on and try to get to your destination, maybe Paterson or somewhere down there. They took a lot of coal. You would try to get down there. You could get unloaded and get back.

Could they unload you on a Sunday?

No. Sometimes.

Sometimes they did?

Not very often.

At least you were there and then they could do it first thing Monday.

Sunday those days you couldn't buy nothing in the stores or nothing like that.

Even the stores were closed. It's kind of hard for us to imagine that today. Did you go to Church along the way?

Lots of times when I was a boy.

Did you eat better on a Sunday than on a weekday or was it just another day as far as eating was concerned?

As far as eating, it was just another day. We didn't make no preparations.

What would be the average day for you? When you got up in the morning, what time would you get up? Supposing you were at a lock and the locktender would say, "Okay, we are ready to pass you through." What time would that be that they would open?

Generally around 6:00.

It varied, I think with the years. Some say 5:30; some say 6:00, but right around there. It was early in the morning, wasn't it? They closed for 8 hours, I understand.

They would close around 7 o'clock, I think.

I understand that they were open about 16 hours a day. Would that be right or isn't that right according to your recollection?

Not according to my recollection.

Maybe about 12?

Yes, about 12 hours a day.

I see. What would be your typical day then? You would have to get the mules up and fed and ready probably before the locktender would start business so you would be ready to go through, so what did you do?

Some of them took care of the mules and took good care of them. They would get up and clean them, brush them, feed them, sort of have their meals before they started out. And some of them would just get up when it was time to go and put a basket on them. Then they had to eat and work.

Eat on the way? Did they mind eating on the way?

They would slow up and you didn't hurry them while they were eating. That is a man that took care of his mules. He would let them take their time. If they must move it was all-right.

Would you eat then at the same time, after you got the mules fed?

Yes.

And then you would proceed on and then what would happen? You'd go through the lock and you would now be on the level.

Often visiting children would assist the lock tender in passing a boat through the lock.

Some mules would go themselves, and some you would have to go out and stay with them and keep them going or they didn't go at all. Some you could let them go and they would go all day. If you got a good mule, boy, that was something for you.

You wouldn't even have to be behind them?

Right. You didn't have to drive them.

Would you put the best mule, the good mule, would you make him the lead mule or the shaft mule?

He was the lead mule.

Because what he did the other mule would have to come along and do also.

He'd have to come along because you had this pulley, see, and you had this pulley and the other mule had to follow to do what the lead mule did.

How often a day did you feed those mules?

At least three times a day.

At least 3 times, maybe 4 though? It was a long day.

When you closed up at night, maybe 12:00, you would give them hay and a good feed. They generally had feed and cut hay. They would mix that up and give them a good meal at night when they went to bed.

Did you have any other relation working on the canal?

I had uncles, aunts, and brothers.

You're quite a canal family then.

Yes.

Mrs. Johnston: Did they work on the canal?

Yeah.

Mrs. Johnston: Your brother Dave?

Yes, he was just a boy when he worked on the canal.

Dave–he's not living yet today?

No, he's dead.

Oh . . . Who were the bosses that you remember? When I say "bosses" I'm not thinking of your father.

"Doc" Piatt was the maintenance boss there at Green's Bridge. Aaron Vough was Assistant Superintendent. Powers was Superintendent.

How about Isaac Durling? Do you remember Isaac Durling?

Durling. What did he do?

He was a boatmaster according to an Almanac I have. A boatmaster's job was to make sure that the boats had their equipment (maybe an extra towline); to make sure that they were outfitted enough to make the trip; make sure the boat was . . . You would say "sea-worthy" if it were an ocean-going ship, you see.

Generally the boatmen had to do that themselves.

Weren't they checked up on that?

No.

They weren't checked up on it. According to an Almanac I have, I think it's 1892, a directory, Isaac Durling was a boatmaster at Phillipsburg.

I didn't know him.

You didn't know him? He's in that picture too. I'll show you his picture.

They had nicknames too.

Yeah, I know. When you were a child working on the canal with your father, did you ever work with any other men?

My father quit boating when I was just a young boy. Then I worked for Luther Dalton.

Was he from Phillipsburg?

Yes. I would get $3 a trip.

How many trips a year would you average about?

He would make a trip in a little over a week. Sometimes he would make it in a week to Paterson and back.

It depends on the day you started, if there wasn't a Sunday in-between, and how good a time you made.

And how everything went, you know.

Sure.

If you struck it all-right where you could make these long levels. Now sometimes you would start out here and you'd get your boat loaded and you'd go out and get on what they called a seven-mile level, that is right here in Washington.

From New Village all the way to Washington.

And you would just get on that thing about 7:00 and go about 7 miles. It would take you about 4 or 5 hours to make that with a loaded boat. You didn't get up there until around 12 o'clock.

Do you remember any toys that the children might have had on the canal? Homemade toys?

No, only a dog.

And yet they did make toys. Maybe I can trigger your memory a little bit. How about whistles made out of willow?

Oh, they did that! They would make them. They would make those willow whistles and they'd blow them, too.

Mrs. Johnston: They had the slingshots.

Did they have slingshots? You could sit in the boat then and pepper the mules a little bit in case they slowed down.

Yeah, they had slingshots.

That is what I mean. They are the homemade toys I mean. Now

little girls on the canal–I understand they made dolls for them. Dried apple dolls or dolls made out of corncobs and they used the silk for their hair. Do you remember ever seeing those?

Yes.

Do you remember any that I haven't mentioned? There is a little toy that someone showed me how to make. It was a shaft with a little propeller on and they would rough one of the tops of the shaft and then with a penny you rubbed it along the edge and this propeller would spin around. Did you ever see those in your canal trips?

No.

Well, I made one and I have it at home and it works very good. But I am trying to get a collection of all the games and toys that they might have used, that children might have used on the canal too. How about jacks?

Oh. They played jacks. And we used to pitch quoits.

Quoits. You mean horseshoes or round ones?

Horseshoes.

Did they have the round iron quoits too?

No. There was only a few, but it was mostly horseshoes. And they used to play cards.

Oh? For money or just for fun?

For a little money, I guess. And then they used to play for drinks.

Yeah, sure–well, that's what I wanted to know. How about buttons on a string?

Yeah, we done that. Pull the buttons and make them spin around.

How about chestnuts? Two chestnuts on a string where one went one way and one the other?

Yes, they did that too. Chestnuts were one of the things.

See, you know these things, but you're not telling me until I trigger your mind a little bit. That's all-right. I just wanted your affirmation that they did have these things.

Mrs. J.: When you say these things, I know I used to do them too.

Right. But one forgets them over the years, you know.

Mrs. J.: I used to play with a button box. I would take the button box out and sort them out in different colors, and I would string them too.

How about checkers?

Yeah, they played checkers too and dominoes.

Would they make cradles, string cradles? You get a big string and hook them around your fingers. Mostly the girls, but I have seen fellows too make little beds like for a cradle.

Yes, they did.

Anything else that I haven't mentioned?

Well, you haven't mentioned how they cooked, did you?

No, I didn't.

They had a little stove that stood about so high and it was round. That's all there was to it. The grate into it, you know, to shake. They just kept that burning all the time. That is where they cooked.

Mrs. J.: In them days they would sleep in the same room.

Now they would cook on the deck too sometimes, wouldn't they?

Yes.

In the summertime it would be too hot in that cabin. And the stove was so small they could pick it up and put it in the hinge deck of the boat. That would be the lowest part where the hinges came together.

Yes.

Then it would clear the bridges much easier, especially if you had a load, is that right?

Yes.

What would be your meals? In the morning, what did you eat and drink?

Some people had bacon and eggs; some had fried potatoes. Whatever you got. Some people didn't hardly have anything.

The boatmen often improvised, as illustrated by the barrel stove in a box of sand at the "hinges."

And coffee?

Oh yes, you had your coffee. You put your coffee right in the pot and *boil* it. They got all the coffee out of the coffee whats in it.

And all you did was keep adding water to it, I suppose.

That's all you had to do was add water. Then you'd drink it boiling. I think it was better coffee than what you get today.

It probably was even though it did get strong maybe. How about pancakes?

Oh, we had pancakes.

Did you call them pancakes or did you call them something else?

Flapjacks.

Flapjacks. And buckwheat cakes? How about dinner—what would be your average dinner?

Well, sometimes you'd make a pot of beans and ham and sometimes you'd cook a little something extra like potatoes.

Mrs. J.: You'd buy vegetables along the way.

You didn't get much meat because there wasn't much. You know you were traveling along there and you couldn't get it. You had no refrigerator; you couldn't save it.

Mrs. J.: No, but you could buy vegetables, couldn't you?

There were canal stores along the way here and there, weren't there? How about supper? What might be your supper on an average day?

Well, anything you got. A piece of bread or something, a cup of coffee sometimes.

Maybe leftovers from what you had for dinner too.

Yes.

All-right, now you are a boy walking along the towpath and it comes dinnertime and you don't have much for dinner, what would you eat then? What would they give you? Would you have time to get on the boat and eat? Supposing you were walking along and they didn't want to stop?

You could get on the boat and let the mules go along to eat.

But sometimes you had a mule that you had to keep after; that was goofing off along the way.

Sometimes you had a sandwich and you'd keep out there with the mules.

How would they get this sandwich to you?

Sometimes you would come to a bridge and when you come to a bridge you are right close and you can come up and jump up on the boat and get your sandwich.

How would you get off again?

At that time you would get right off.

Right at the same bridge? Did they ever wrap it in newspaper and throw it at you?

I never had any like that.

Somebody told me once that they got this newspaper and they'd wrap the sandwich up and they would throw it over the towpath to them, especially if there was no other way–no bridge. Was there another way you could get off that boat and on again without using the bridge?

Oh yeah. They used the pole. A guy could put the pole, say,

here, and I could jump out to the street.

You could swing out on the pole like a pole vault. Only you didn't go over the top, you would swing to the side, like.

Yeah.

What would happen to the pole?

Nothin'.

I mean, you wouldn't leave it sticking in the mud or the canal. Who would pick it up? Would they pull that out from the boat?

No, the towpath boy picked it up himself and handed it back to them on the boat.

And then you would walk up to where the mules were?

See, the boat wasn't any further than from here to that wall away from you at any time.

Yes, from the side of it, but you were about 150 feet away from it when you were out there with the mules, weren't you?

Well, yes.

Because that line was long. How long was that line, would you say about?

Oh, I don't know. About 100 feet long.

Maybe a little more or a little less, but right around that average. How about fighting on the canal? Was there much fighting or is that exaggerated?

There wasn't too much. There were some gangs, bad gangs, down around Newark. They would ask you to give them a nickel or a dime and you had better give it to them or they would remember you.

Little gangsters, huh?

They didn't try to bother one another.

But how about among the boatmen themselves? Was there much rivalry there? Was there much fighting among the boatmen or was that exaggerated when they talked about it?

No, that was exaggerated. They were all nice and friendly with one another.

But they tell me that Art Unangst was quite a fighter though.

He used to fight for the fun of it. Is that what you found out?

Yes.

Not to be nasty or mean, but he just liked a good scrap.

He could tell when somebody was pretty good with their fists, then he would go challenge them to see which one was the best. He was pretty good, Art was, but he'd get punched out too every once in awhile.

Yeah, I guess you can always find someone who can clean your clock. But, he wasn't a real big man was he–Art Unangst?

He was big enough, but he wasn't a 6-footer or anything like that. No, but he was a pretty good sized man though. I remember that. You can go by his brother there. You see him–Henry. Then there was another one–Carm.

And Fred.

I didn't know him.

There was quite a family of Unangsts. Ozzie. Well, we have covered quite a bit. How long did you work on the canal? When did you leave it?

I left when I was about 20 years old.

You never went back again?

No, it didn't run much longer then.

It was pretty close to the end of it then?

But I used to have a good time. I used to always enjoy the canal. You were all the time on the go, see. You would be in Stewartsville, next New Village, next you would be in Washington, and Port Colden and Hackettstown and you went all over. And you knowed everybody and everybody would wave at you.

Mrs. J.: You used to stop in New Village and go to a dance or something like that.

No, I worked on the mud-digger and I had a girl in New Village.

When you worked on the mud-digger, sometimes would you go out for an extended length of time, maybe a week?

Oh, we would go out for a month.

You would go out for a month before you would get back to

Phillipsburg?

Yes. Sometimes you would go out away from Phillipsburg and you wouldn't get back until fall.

Is that right?

You would travel and sometimes there would be a hard shower and it would wash the mud in the canal and you would be rushed there to dig that out quick. Then you would sometimes work there maybe a month and you would move on to some other place. You was kept busy.

Now there were two mud-diggers. One was stationed in Phillipsburg and one probably down by Jersey City. Do you remember any of the other crewmen on those diggers at various times?

I know Kugler was the one that run the other digger, but I don't know any of the other fellows.

How about the one on this digger here that operated out of Phillipsburg?

I knew them.

Could you name them? How about Amos VanSickle? Did he work on it?

He worked on it too.

And Art LaFever?

Harry Van Sickle, and Bill Gross. Oh, I can think of them, but I cannot think of their names.

How about LaFever?

LaFever, he worked on there for years.

What was his full name?

Johnny.

John LaFever.

He lived on there the year around. His home was in Ledgewood, but he stayed right on the digger. It was cheaper, I guess.

How about Teddy Dailey?

Teddy Dailey worked on there for awhile.

Teddy was also a cook on there at one time. Do you remember when he was a cook?

Yes. He was a cook when I was on there. He didn't stay too long.

He told me that if you gave him an old piece of hambone, he could make soup and it would last for a week. All he had to do was add more water to it.

That's true enough. We lived on ham and soup and stuff.

And cabbage? He said he used to bake potatoes and hand them out to the men on the shovel while they were still hot.

Yes.

He said one day the fog was so thick, he said "Talk about fog! I worked on the Morris Canal and the fog was so thick that I didn't see land for three days!" And you were only 8 or 9 feet from land so it must have been pretty thick. I think he might have been stretching it a little bit, but it is kind of hard to imagine somebody being on the canal and not seeing land for three days. But I guess it did get foggy. Do you remember anything exceptionally funny on the canal? Some misfortune happen to somebody that was exceedingly funny that you thought about the rest of your days or do you remember anything that was out of line or funny?

A morning mist or fog frequently covered the canal and countryside through which it passed. Location: Newark.

Nothing out of line. I always thought they were a nice friendly bunch of people. Everybody knew one another and talked to one another. You could go away and leave your cabin unlocked—nobody bothered you. You could lay down on top of the cabin if you wanted to. The mosquitoes would get you at night. There were no mosquitoes at this end, but after you got up to Guinea Hollow Dam, oh, the mosquitoes were terrible from there all the way down.

What would a boy do on the canal? Where would he sleep at night in the summertime?

Down in the cabin.

Would he ever go sleep with the mules?

Oh, sometimes he would. But there were more mosquitoes there in the stable than any other place.

And you wouldn't sleep on the boat itself, on top?

Sometimes.

Sometimes you would. Depending on where you were, I suppose.

You could get under the deck and lay in the coal heap any old way.

Now that was a very small cabin. Some people raised their families in the cabin. Some had girls and some had boys. When it came time to go to bed where would the girls sleep as to opposed to where the boys would sleep?

Well, they did try to keep them separate. Some layed on the floor, but there were only two beds.

Only two bunks?

Only two bunks.

Did they swing down from the wall?

Yes.

And in the daytime, they would be put back up again—is that it?

Right.

How about in back of the cabin? Was there a little door that you could crawl back in there in the bilge of the cabin; Was there a little space back there that you could get into?

There was a space where you steered the boat under there. There was a place where you had a bed.

You had a bed there too?

But there was only two. That's all.

Oh, that was two. There was no false work where you got between the boat and the cabin? There was no other space in there?

No.

Not in any boats you remember?

There were square boats you know. Then they made them with a round stern. Then they had two beds, one on the bottom of the floor that would swing down and one up above that would swing down. You see with that "round" you could not get no bed back in there.

Which were the oldest, the round or the square?

The square boats were the oldest.

Then in later years they made them round. What was their advantage? Why did they go from square to round?

I never could figure that out.

It seemed it would be harder to make them round. A little more work.

Someone said they were easier to steer, I don't know.

Mrs. J.: Perhaps the square ones would poke into the side of the canal and it would be harder to get going.

There must have been a reason for it, but they stayed with those square ones for a long time and all of a sudden when the canal is about ready to finish, then they made some round ones.

They thought that they pulled easier and had the advantage of steering nicer. I think they did too.

Now those with a round stern (Because that's where the cabin is located.) there the bed was. . . .How did you say?

There wasn't no bed in the back. They were on the side.

Both on the same side.

Yes, one above the other. Then you had a table in the center.

Was that hinged too or did that stay up all the time?

It was and when you wanted to use it, you would raise it up and put a stick that come up under it.

What else might be in the cabin?

That's all.

Would there be a counter maybe??

Oh yes, you might have a counter.

How about a lamp?

You had a kerosene lamp.

All-right. There would be little windows in there, wouldn't there? Would there be curtains at those windows sometimes?

Sometimes.

If there was a woman along, then there might be.

Mostly they didn't.

Mrs. J.: They used to have lanterns on the canal boats over in the Delaware.

Mostly all lanterns, yes.

How about a corner cupboard? Would there be a little corner cupboard to hold the dishes, maybe?

Oh, yes, you had just a little corner cupboard—that was all.

Mrs. J.: There was nothing else there.

And you had a stove.

A stool? You had only the one chair or was there more than one?

I said stove.

Oh, a stove, I'm sorry.

They had two chairs, not very big chairs.

How about a wash basin and dishpan?

You'd either wash in a wash basin or you'd go out and wash in the canal.

Yeah, but you had a dishpan probably.

Yes, we had a dishpan to wash dishes.

Where would you store your food? You must have had some food, some cans, some preserves. Would you have a shelf

where you might have that out of the way?

Yes, there was a shelf in there and a lot of them used to put like their bacon and stuff like that in the oats.

They used that as a refrigerator. The feedbox. I have heard that so many times that they would put it in the feedbox. But that wouldn't be the fresh meat. That would be the smoked meat, wouldn't it?

Fresh meat you couldn't carry. If you got that, you would have to cook it right away.

Right. Was salt mackerel a favorite dish on the canal?

That was the favorite dish.

How about potato soup? Was that made quite a bit?

The mainstay was potatoes and mackerel and they'd cook a pot of beans and buy a chicken and cook a chicken and make a lot of soup.

Art Unangst was telling me about one day he went to steal a farmer's chicken and the farmer came out and caught him and said, "I see you catching my chicken!" And Art said, "No, this is mine. He got off the boat. I am trying to catch it." So the farmer helped him catch his own chicken and gave it back to him. So they were pretty foxy, weren't they?

Mrs. J.: Yes.

Yeah. The boatmen, they all used to have a barrel and they would bury it in the ground. They would put cabbage and turnips and potatoes in there. And then they would have them full for the winter. They would have their money, but at Christmastime, they got about $40 or $50 back money that they kept away; they kept out for them. And they lived on that the whole winter long. They bought cabbage; some of them had a couple of pigs. They rented houses, you know, some of them did. They only paid $2 or $3 a month for a house. Then you would buy coal for $2 or $3 a ton. They really lived and lived nice.

Not real good, but they knew where their next meal was coming from.

Yes, they knew where their next meal was coming from.

Some of them would have jobs in the wintertime too, wouldn't they?

Some of them would work; mostly ice. They didn't have refrigeration, you know. The business was quite a business when I was a boy. The ice wagon would come around and you could buy ice. You could put that in your refrigerator and that would keep things cold for you. That's what they had.

I know Pete Lenstrohm used to cut ice (up by the island they would get it) and cut it and float it down to Hagerty's where the lumberyard was there. And then they would bring it up on a conveyor. Did you ever do that or see them do that?

I have seen them do it.

That must have been quite a trick riding those cakes of ice down the river through those rapids down there over by the horseshoe works and everything.

Mrs. J.: You don't see real thick ice like that anymore.

No, you don't seem to. I agree with you.

You would see them cut ice there right from the old bridge on the Delaware River and then a lot of people had little ice-houses and they were located around the canal. It was quite a business.

Yes, it sure was.

They had everybody cut ice. And then in the summertime, they would take it around and sell it to houses.

So that would supplement their income too.

Yes.

Now, they had a company store down here in Phillipsburg. Do you ever remember being in that store?

Yes.

What was in that store?

They had everything you wanted in the can line, canned food.

And also things you might use on the boat like washtubs, or basins or mops?

Mostly everything like that.

Did they pay cash for it?

The company owned a store and stable at Port Delaware at Phillipsburg. The stable was a principal place to buy mules.

They didn't have to.

They didn't have to?

They would take it out of your check every time you came. They would take it out of your hide too, I guess.

I guess it wasn't cheap, was it?

No. They had mules there. They had a big stable and they would sell mules.

Yes, they had the mule stable right in back of it to one side.

That's right.

How much did it cost generally to stable a mule overnight?

A quarter.

A quarter. 25¢. That would entitle you to do what? Just for the stabling or would it give you hay?

It would give you hay.

Oh, but no oats or anything like that.

No.

Where did you buy your oats?

From the stores along the canal. It used to be like every lock had a store. Well, not every lock, but like down to Green's Bridge like around Green's Bridge and Willever had a store.

Then there were little stores all the way along. Up to #10 Plane there was a store.

But they didn't all have oats, but you knew far enough in advance when you were getting low so you could buy oats along the way.

Mrs. J.: And plenty of tobacco, chewing tobacco.

Oh yes, chewing tobacco.

How much would a mule cost you a day? Now here we established 25¢ for overnight and how much in dollars and cents would it cost you to feed them throughout the day on the average? Another quarter to 50¢ or something like that?

Yes. It didn't cost you too much. It didn't dare to.

Yeah, that would be about 50¢-75¢ a day for one mule. So then that would be about $1.25 to $1.50 for two mules. Or was that too high for the two mules?

That was too high. I don't think it went that high.

That is too high. Then 25¢ might have been the stabling for two

Basins or widewaters allowed boats to turn around, pull aside to make repairs without blocking traffic, or get supplies. Location: stores near Stanhope.

mules.

Yes, because you only got about $40-$45 per trip.

That is what I am trying to get down to. So it would cost you about 75¢-$1 a day for the care of your mules. That might be a better average. When you said 25¢ per night for a mule you must have meant for the two mules and I took it for one. Because they didn't have these stables to make money. I guess they were for accommodations for the boatmen more than anything.

Then they had a store there too and you would buy something off them.

How about some of these lockhouses? Would the wife of the locktender make baked goods sometimes and sell them?

Yeah, you could buy homemade bread–three loaves for a quarter. I remember that.

That's cheap. I'd like to buy some today for that. I just paid 57¢ for a loaf of bread.

And that was good too.

Do you remember any lock called "Fresh Breads" Lock?

Yeah.

Oh, you do? Where was that located at?

That was located down this side of Dover. "Fresh Breads" they called it.

Would that be Ledgewood or beyond Ledgewood? Before you'd get to Port Oram?

Just as you went into Dover, beyond Ledgewood.

All-right. Now Port Oram came first. You went past Port Oram?

Port Oram?

Yes, that was Wharton in later years.

Wharton. Yes.

That was before you got to Wharton?

Yes.

That was at the foot of #4 Plane. There was a stone lockhouse. Is that where you mean?

Yes, that's where I mean.

Major transshipping locations along the canal were called ports. Most faded in importance. Location: Lock 2 East, near Port Oram, now Wharton.

That was Burd's Lock. A family by the name of Burd ran that for many many years. I don't know who might have been there when you went through. So is that the place you mean?

Yes.

"Fresh Breads" Lock. They also made bread out here in New Village and it had the price list up on it, "Fresh Bread", up at Gardner's Lock, you know.

Oh.

They had pies and they had a small store there.

Yes, they had stuff there. They had a little store.

That's right out here in New Village.

But "Fresh Breads" Lock was down there before you go into Wharton?

Yes.

Do you remember any songs or ditties or parodies that were sung on the canal? I don't mean "You rusty ole canaler, you'll never get rich working in the canal, you s.o.b.", you know?

That's what they used to call it.

That was a very common one, but it wasn't bad.

"You rusty canaler, you'll never get rich", and then they would swear about it, you know.

"You son of a bitch, you're working your life in a ditch." Yeah. Do you remember any other songs? There was quite a bit of

singing on the canal they tell me.

Yes, I know. I just can't now recall. It'll come to me, but I cannot recall now.

I would like to record as many songs and ditties, especially pertaining to the canal as I can. Do you remember a song, "Go along mule, don't you roll those eyes, you can change a fool, but a doggone mule is a mule until he dies." Do you remember that?

No.

I'm not sure if that was a popular song of the day that the canal men adopted or if it was one that somebody made up. But there was another one "Going down to Cooper's". This applied only to the Morris Canal and I will play that for you. What else do you remember about the Morris Canal that I have not asked you about?

Oh, there must be something. It's too much at once.

Yeah, I know. Did your father ever tell you any interesting stories about when he started on the canal? Experiences that might have happened to him that you remember?

Not that I recall.

Did he ever tell you about the big cities? Going down to Jersey City looking across the river and seeing the big buildings? Did that more or less make you want to travel on too? You know people didn't go very far in those days. If they went 10 miles from home, it was quite a trip, but here you got in this canal boat and you went 100 miles. It was quite a thing.

Well, they used to go down and go on to the Passaic River.

Yes, we talked about that before and it's a lot clearer to me now. When you got down to Newark and went down the inclined plane and you went under the city market, that was a pretty tough place down there, wasn't it?

Oh, boy! You would go under there and you had to push the boat through with a pole, you know. And that market, that would be under a place where they had a hatch. They would open up this thing and dump this garbage. Sometimes it would strike you right in the face. Oh, it was tough when you stop to think of some of that stuff.

The eastern end of the tunnel under Newark's Centre Market. Sign reads: "NOTICE. DUMP NO GARBAGE ON CANAL PROPERTY."

How about rats? Did you see rats running around under there too sometimes? And water dripping down?

You couldn't see no rats because it was all walled up. It was dark. There was no ground or anything.

And then you went down a little further to Mulberry Street and there you came to another lock. Now what were the distinguishing features about this lock?

That lock was called the deep lock. Oh, it was down I imagine about 75 feet.

Well, it raised and lowered the boats 20 feet altogether, but the lock might have been a lot deeper where the water was. But the change in elevation in that lock was 20 feet and that was pretty high for a lock. You know generally when you went that high you would go over an inclined plane. Then you could go directly into the Passaic River if you wanted to or you could turn (You had a "Y" there. Is that right?) south and go down the river aways and then cross over into Jersey City. Right so far?

Right.

Now there is a chemical plant down there between Mulberry Street and where you crossed over. Do you remember anything at all about that section down there? If you ever got any loads or took coal to that chemical plant or got any loads out of there or anything?

No, I don't. I know we went down pretty near Jersey City. It could have been.

Yes. Now when you crossed over, you crossed over by the bridge there. You crossed over the Passaic River and then you went through a piece of land and then you went in to the Hackensack River and then you went into Jersey City by the pump house. So far that's right.

Yes.

How about the tides? That would affect your crossing, wouldn't it?

That affected your crossing. But they used to have a pump there to keep that (that thing was an eight-mile level they called it) where they would pump water out of the Hackensack River into there to keep the water in the canal.

That's right, but how about crossing over there now—the tides. You couldn't cross at low tide, is that right? And high tide was the best time to cross. If the tide was ebbing or flowing, it was a little tricky sometimes, I imagine. Otherwise the tide would swing the boat way out or swing it against the bridge, wouldn't it? So you had to be very careful about crossing there.

You had to know your job to do it.

Yes. Right. And the mules would walk up on the bridge. There was an extension that was built out on this bridge just for the mules. Is that right?

That's right.

And then you got over to the Jersey City side and there was the pumphouse. And then you went down the Newark Bay into Bayonne. Then from Bayonne you went to Greenville and then from Greenville you went back into Jersey City again. You'd hit Jersey City twice. You'd hit West Jersey City and

then you left it and skirted around and came back into East Jersey City. That was always an interesting crossing for me. But sometimes your loads took you up the Passaic River and then when you left Mulberry Street you would go directly into the Passaic River. Was there a guard lock there that let you in to the river?

Yeah.

I would imagine there would have to be. And then you wouldn't go into the river though until you had your tugboat pretty close by, I suppose.

You would put your mules in the stable and they would keep them for you and put feed there for them until you came back. Sometimes you would be gone for 2 or 3 days.

Yes. Where was the stable located? Was that by the lock at Mulberry Street or by the river?

By the lock at the river.

Oh. Now when you came to the top of the plane in Newark (that was plane #12 East), how would you take those mules down? You didn't put them on the boat and there was no towpath down there. Where would you walk the mules to meet them?

You had to walk around through the city to get there.

Through the city of Newark. Did you ever run into any trouble walking those mules through the city? Any smart kids ever challenge you or anything?

No. Never no trouble. I never had any.

You had two mules to take care of and you walked right in the streets. Naturally you didn't walk on the sidewalks. But the first time you did it, how would you know where to go?

Sometimes the mules could take you.

Just let them go themselves?

Let them go and you go with them and they will take you right through.

Do you remember on the canal any canal preachers or missionaries that would come and preach at Port Delaware?

Yes, I do.

What was his name?

My wife can tell you. Hey, Eva. What was the preacher's name that lived on Washington Street and used to go down to the boys in Port Delaware?

Mrs. J.: Williston.

Francis Williston. Were there any others that you know of?

That was the only place that the boys got any literature given to them.

But was there another man that used to come down and talk maybe? Did you ever hear of a Reverend Cook from Brainard Street?

No. But this here little man, Francis Williston, he came down there and he was there every Sunday to give to these little

Second from the right is Rev. Francis Williston, who with his assistants used Rush's printery in Easton, PA, to produce religious tracts.

boys on the boat. And that was nice. I'll tell you that helped me a whole lot.

Would he give you a little card or something with flowers on it?

Yes, and he talked about the Lord. And it was just something that kept me thinking. I'll never forget it.

Do you know that that same man went over to Rush's Printery on Wood Street in Easton and on his own time, he used to print these things up and save expenses and money?

He did? I didn't know.

I have a picture of him in the Printery working away there with the boys. Printing these tracts up so they wouldn't cost him so much money.

He was certainly a nice old man to these poor boys, I'll tell you.

How about the S.P.C.A.? Did they ever check your mules when you went through?

They did down around Newark. They would come there and look at your mules and raise their collar up to see if they were sore. They would have you arrested if need be. You had to take that collar off and put what you call a press collar on below the shoulders until the sores healed up. Then they had to be covered up so the flies couldn't get at them.

Yeah. Was it mostly at Newark that you did this or were there other places along the way?

Mostly down around Newark.

The small towns generally couldn't provide for this.

They did in Newark.

I guess most people took care of their mules anyhow, didn't they? Because after all, that was their livelihood. If they didn't have mules, they weren't going to go anywhere.

Well, I think we are pretty well finished up. The only thing we don't have is a song that you remember that you are going to remember.

I will try to remember that. I know we used to all sing it. Once in awhile I am humming what I can remember of it.

Yeah. Well, when you think of it, write it down right away.

I will.

Plane 12 East at Newark was one of three double track planes on the canal.

JOHN HENRY LEWIS MUTCHLER

Born at Shillinger's Mill near Cooksville in 1857. Died in 1933. Worked as a driver, on the maintenance crew, and for the Morris Canal Abandonment Authority.

Chapter Ten

John Henry Lewis Mutcher

John Mutchler was born June 23, 1857 near Shillinger's Mill at Cooksville, now known as Stewartsville, New Jersey.

His father, John Mutchler, was a section foreman on the Morris Canal stationed at Broadway, New Jersey. His uncle, George M. Mutchler, was a foreman and later a supervisor on the canal.

At the age of ten, John H. L. Mutchler started work on the canal as a mule boy, driving the mules on the towpath. Several years later, he was promoted to the work scow with his father at Broadway. The latter was foreman at Broadway from 1857 to 1888.

In 1884, the canal company thought John was getting along quite well, so they sent him to Hackettstown, New Jersey as a foreman to succeed Jacob Shields. He spent seven years there. He returned to Broadway as section foreman on his father's retirement April 16, 1888. His father passed away the following month, May 10, 1888.

John always said that the men who followed the canal never made too much money. In the early days of the canal, the men employed worked for $1.50 for a ten or twelve hour day. During the panic of 1873 and continuing until 1877, they had to work for as low as 80 cents a day. Work was so scarce during that period that carpenters and masons came to the Canal Company and begged for work shoveling mud out of the canal at that wage.

Low wages were the rule. Foremen were paid only $45.00 a month and this meant for seven days a week. They had six or eight men under them, but in 1888, the call for economy sounded and the force was cut down to two men. This situation prevailed till just before World War I when the government took over the railroads and canals. Prodigality was the rule and the canal men, along with the railroad men, got substantial raises—fifty cents an hour.

Often the scow would bring loads of dirt to help repair a breach. Dirt was available at several planes. Location: Plane 7 West, Washington.

The Morris Canal was for a great many years operated and owned by the Lehigh Valley Railroad Company. Unfortunately, the Railroad Company did not include its canal employees in its pension fund.

When the canal was officially abandoned by the State of New Jersey in 1924, Mr. Mutchler worked for a time with the Morris Canal Abandonment Authority.

Then in 1926, after fifty-nine years with the Canal Company and reaching the age of sixty-nine–a time when most men have thoughts of retiring–John Mutchler went to work for the New Jersey Interurban Coach Company as a repair man and mechanic.

On October 2, 1928, Mr. Mutchler and his wife, the former Carrie Willever of Broadway, celebrated their fiftieth wedding anniversary.

Mr. Mutchler worked up to the time of his death which occurred in 1933. He died of paralysis after one week of illness.

He is buried in the Stewartsville Presbyterian Cemetery, Stewartsville, New Jersey.

John H. Mutchler with his favorite mule. A protective covering occasionally was used in place of a fly net.

LEON LAKE

Age 86 in 1976. Born at Phillipsburg in 1890. Worked as a section hand. Now living in Easton.

Chapter Eleven

Leon Lake

"We done some pick and shovel work. We moved the track on the plane; repaired the track. Where anything needed repairing, we did it just to get the canal in shape."

The following is a conversation taped of Mr. James Lee interviewing Mr. Leon Lake:

Now what is your full name?
Leon David Lake.

What was your connection with the Morris Canal?
Well, I was hired on for one month for just repairs. I enjoyed working because it was a good little job. They said it would last about a month and it did. And it was a wonderful place to work. I started from Phillipsburg where they repaired the boats and worked all the way up to Stewartsville.

What actually did you do?
We done some pick and shovel work. We moved the track on the plane; repaired the track. Where anything needed repairing, we did it just to get the canal in shape.

And what month was that?
Well, I would say it was early spring. It was April, around that time.

How many men were in your gang?
I would say there were about eight men.

Do you remember what year that was?
1912.

How old are you now, Mr. Lake?
I'll be 86 this coming December.

Did you ever dig out the mud around the cables?
We dug out the mud around the cables and done some repair work with hammer and nails, and helped move the tracks and make them fast—everything to get ready for spring.

How many hours did you work a day, as far as you can remember?

Well, I remember it was nine or ten hours a day at that time.

Do you remember how much you got a day?

I think it was $1.25.

And how many days a week?

Five days a week. We didn't work Saturday.

Did you come in contact with any of the people living in the plane houses along the way?

I came in contact with all of them, but the funny part is I try to remember the names, but I forgot the names. But I got in contact with all of them.

Did you work on the plane houses at all where the machinery was?

Yeah, I worked on some of them, too.

Do you remember what you might have done?

Heavy rains or hard winters caused silt to build up at the base of planes. Crews removed it by hand. Location: Plane 5 East, Dover.

Well, anything that needed fixing, we fixed it—that's all. Loose boards or something—we done it.

You never had any relation working on the canal?

No, I didn't. I just happened to take the job as a young fellow and I enjoyed working on the canal at the time I worked.

Did they have the canal drained of water or just lowered?

It was drained of water pretty well.

The other men you worked with—were they temporary workers or permanent workers?

Two or three were temporary workers. The others were permanent workers on the canal.

They just supplemented their force when they hired you?

You'd always see them downtown to talk to them at that time. Years ago, I knew their names, but right now, just because I want to recall their names, I can't.

Maybe you will later on.

I would know them again if I see them.

How long did you have to wait for your money?

One month. We were paid by the month at that time.

By check or cash?

Check, as far as I can remember.

Did you hear any stories while you were working there?

Oh yeah. We talked about local things going on, about the canal, and about the things going on at the time I was working.

Did they push you very hard or could you go at your own pace?

No, just ordinary working. No pushing at all. I had a very good boss.

Did you do any singing while you were working that month?

Well, I'm not a singer, but I heard them singing, as I worked along.

Do you recall any of the songs that they might have sung?

They must have been the fellows that were steady workers.

Do you recall the songs?

Well, all the old-time songs. I know there was one that sounded like a canal song. I enjoyed hearing them.

You don't remember the words to them?

No, but they were the old-time songs of that time. "Bicycle Built for Two" was one of them.

The songs that were popular for other people were also popular to the canalers.

Yeah.

How about your lunch. Did you take your own lunch or did they feed you?

Oh, I took my own lunch. I packed my own lunch. I enjoyed working there. They said it was just temporary and I was young and I just enjoyed taking it, that's all. I admired canal life because they go along the towpath so slow and easy, driving them mules and sitting on the boat. Some had homes and some lived on them.

How did you get from one place to another? Did you walk all the way?

We just worked as we go along. Then going back, they'd have a horse and buggy to take us back to Phillipsburg again.

You don't remember your immediate boss though?

No, I don't. I got a picture where they repair boats down in Phillipsburg. I used to go down there and watch them even before I ever took a job on the canal. I'd watch them caulk the boats with the stuff they used to repair the cracks in the boats with. They had a special hammer, you know. I used to enjoy that. See, I just enjoyed the canal. In fact, I guess I would have liked to be a canal worker, but my relation never had that. All my folks were engineers on the Railroad, and I worked on the railroad myself. I was on the Valley nine months and fired up on the Valley. But if I'd had relation on the canal, that's where I'd alanded because I liked the looks of the life of it.

Now you know I'm taping this, Mr. Lake. Do you have any objections to me taping this to let other people hear?

Oh, no. Not at all.

Chapter Twelve

The Morris Canal in 1880 by John Johnston

After the discovery of anthracite coal by a hunter named Philip Ginter, on top of Sharps Mountain, now Summit Hill, nine miles west of Mauch Chunk, Pennsylvania, in the fall of the year 1791, the private and public demonstration of the stone coal, as it was then called, proved it to be such a useful fuel that a new spirit was awakened in the destiny of the people.

Then the problem of how to distribute the coal by means of transportation was one that remained for the future to solve.

With a rough, rugged country and a few miserable winding roadways, the matter of transportation was a deeper mystery than ever.

But finally after many years of consideration and planning, the only available way to market the coal was by boats, and to this end a canal was constructed along the Lehigh River, by means of a series of dams in the river and locks in the canal. The boats loaded with coal were brought down along the Lehigh River to the level of the Delaware River at Easton, Pennsylvania.

Across the river from Easton to the east bank of the Delaware lies the town of Phillipsburg, New Jersey. In the year 1810, coal was supplied here, but the places that needed the coal the most were Newark, Jersey City, and New York City.

These towns were about seventy miles away, across the State of New Jersey, by turnpike road through a rough rugged country with its hills and valleys. The transportation of coal over this route was next to an impossibility. In the year 1818, a man by the name of George P. McCullough, of Morristown, New Jersey, while with a fishing party at Lake Hopatcong, conceived the plan of constructing a waterway from Phillipsburg to Jersey City, using the waters of the lake to feed the waterway. In this way, coal could be shipped by boat from

Mauch Chunk, Pennsylvania, to Phillipsburg and through New Jersey to tide water and the Metropolis.

With such an idea in mind, Mr. McCullough consulted an English engineer named James Renwick, then professor of Natural and Experimental Philosophy at Columbia College.

Mr. Renwick took up the work of making the survey and supervising the construction of the waterway, and Major Ephraim Beach selected the route.

On December 31, 1824 a charter was granted to the Morris Canal and Banking Company and work was commenced on building the Canal.

The route selected was Jersey City to Hackensack, then to Plank Road, on to Newark, Bloomfield, Beavertown, Montville, Boonton, Parville, Denville, Rockaway, Dover, Sam Heaton, Drakesville and Powerville. Here the canal reached its summit to the height of 914 feet above sea level. To reach this height, the canal went through 24 locks and over 12 inclined planes.

The Morris Canal was the only canal in the world that had hydraulic inclined planes to raise and lower boats from one level to another.

Then from Port Morris down to Stanhope to Waterloo, Guinea Hollow Dam, Port Murray, Port Colden, Washington, New Village, Stewartsville, Hardport #10 Plane West, Green's Bridge, and Port Delaware, to the level of the Delaware River at Phillipsburg, New Jersey, a distance through 10 locks and over 11 planes of 760 feet, making a total rise and fall of 1,674 feet in a distance of 103 miles, although the air line distance was but 55 miles.

The plane cars were in two sections with hinge bars for couplings, so as to allow the car to bend at the summit of the incline plane. That is, the plane and car were hauled out of the water of the upper level, up a short incline to the summit, then down the long incline to the water in the lower level.

Each section of the plane car was built of heavy oak timbers, and of framework design. The chassis of the car consisted of cross beams about eight inches square, and eight feet long, connected to an eight inch by ten inch square side beam with a mortice and tenon joint securely fastened with wooden pins. These side beams were about forty feet long, and at each end resting on a two-wheel truck of tandem fashion. Each section of the car had four trucks to carry it up and down the incline. Each truck was independent of the other and ran on a track of about a twelve-foot six inch gauge. The frame of the truck was carried on the wheel axles that extended about twelve inches on both sides of the wheel with a heavy brass bearing held firmly in the heavy yoke. The wheels were of a two-flange type—and wide enough to run on a rail of about four inch cap making it more of a shove wheel than a track wheel.

Plane cars as well as lock gates required heavy timbers. Location: Plane 2 East, Ledgewood.

The canal ran under 259 bridges and over 45 culverts, with 43 overflows and spillways. Some of the bridges were private and some public. Some were change bridges. They were used to cross the Canal where the towpath changed from one side of the Canal to the opposite side of the Canal.

The Company built their own boats at their own boatyards. The boats were a two-box type, and bow box and the stern box. Each box held forty tons of coal and were connected together with heavy iron hinges which allowed them to bend with the plane car when they went over the plane. There also was a beer boat and a pay boat that ran the canal.

About 900 boats were used in 1866 to haul the cargos and 1300 mules furnished the pulling power. Ore, pig iron, lumber, flour and feed were also hauled over the waterway.

There was a small country store at every plane and lock that supplied the necessary provisions for the boatmen. Some of the stores added a boat stable to their business that provided oats and hay for the teams. The boatmen always lived good on their boats. Many boatmen had their wives and families on board during the boating season, while others had their wives and families live on their boats all the year round. They did their cooking by means of a small cylinder stove securely fastened to the deck of the boats at the hinges. Here is where the feed boxes were placed that contained a large oat box.

The oat box was the "electric refrigerator" for the boatmen. They kept all their provisions in the oat box, covered over with the oats, which gave the provisions a very delicious flavor that could not be obtained by any modern appliances. The cabin at the stern of the boat was very small, it contained two drop bunks and a drop table plus a corner cabinet.

Because of the consecutive curves in the waterway, it required someone at the tiller handle, which operated the rudder blade, all the while the boat was being drawn by the mules. This was a more strenuous occupation than it appeared to be.

All planes and locks ceased to operate at 9:30 p.m. on Saturday evening, and stayed closed until Monday at 5:30 a.m.

As a boat cleared a plane or a lock at 9 p.m. they ran on until they reached the next plane or lock at the end of the level, regardless of time.

Each plane had an average 9% grade and rose from 35 to 100 feet. It took two men to operate a plane, one man on the plane-car and one man in the plane house tower. He lived in a company-owned house and received $30.00 per month; $100 per month plus house in 1923.

Had all locks been used with their limited lift on this waterway, it would require between 200 and 300 locks to do the raising and lowering of the boats.

LOCKS

All the locks were of the head gate and wicket gate type. That is, to fill the lock with water, the lower head gates and lower wicket gates were closed, then the upper wicket gates placed at the bottom of the lock were opened, and through these gates the water rushed with such force that in a period of

Looking at the drop gate of a composite (stone) lock with additional timbers which served as buffers for the boats.

about ten minutes the lock, which was ninety feet long, eleven feet wide and ten feet deep, was filled with water from the upper level, and to the height of the upper level.

Then the upper head gate was lowered, and the boat drawn into the lock with the boat launched nicely in the lock. The upper head gates were raised and the upper wicket gates closed so as to allow no more water from the upper level. Then the lower wicket gates were opened, and in approximately the same period of time, the water rushed out of the lock into the lower level, and the lower gates were opened, the boat drawn out of the lock and along the lower level until the next lock or plane was reached.

The locks were all a one-man job. He lived in the company's house located on the bank of the lock with rent free, and coal from the boats, and $30.00 per month; $75.00 per month in 1923. He attended to all the wicket gates and head gates from 5:30 a.m. to 9:30 p.m., six days a week while the boating months were open.

PLANES

The twelve planes east and the eleven planes west of the summit were all operated with their own water power. Their incline had a 12% grade and varied in height from 35 to 100 feet.

The water was taken from the upper level of the canal, through a raceway, to the penstock, in the plane house tower, down through the penstock about five feet in diameter, and a fall of forty or fifty feet to an undershot water turbine, which was connected by a bevel gear to a large iron drum. This drum or cylinder, as it was sometimes called, was about nine feet in diameter, and on it was wound a wire cable two inches thick that hauled the plane cars up and down the incline plane. This drum was placed at the same level as the bottom of the canal of the upper level. Both ends of the wire cable ran towards the upper level at the bottom of the canal in the upper level, and were placed on two large shove or groove wheels

A Jersey team consisted of a white mule and a dark mule. Note: Canallers wear hats for protection from the sun. Location: Plane 8 West, Stewartsville.

about fourteen feet in diameter lying flat in a yoke firmly secured in the desired position. One end of this wire cable went around one of the shove wheels and attached to the upper end of the plane car. The other end of the wire rope ran around the other shove wheel, out of the water of the upper level over the summit and down the inclined plane carried on ten-inch diameter idler pulleys, held in an iron stand fifteen inches high, and placed on the outside of the car track to the large shove wheel placed at the bottom of the lower level, thence up the incline on idler pulleys of the same size and make, held in the same way, placed in the center of the car track and attached to the lower level end of the plane car.

Sunday was a day that was very much appreciated. It gave the women a day in which to do the family wash, and the men a chance to look over their teams of mules and to make repairs to their towlines and snubbing lines by resplicing them.

After President Abraham Lincoln's Proclamation, some colored folks came north and engaged in the occupation of boating as a means of livelihood. They settled in an area by themselves just outside of Washington, New Jersey, along the

The inspection team ate well, traveling with a full time cook aboard. Fishing was a diversion, rather than necessity. Location: between Hackettstown and Saxton Falls.

Canal, and because of a sharp turn in the Canal, the site was known as the Black Crook. Every summer the colored folks always held their camp meetings at this point. The second Sunday of August was usually the date for their camp meetings. Three shows were always given–forenoon, afternoon and evening, along with services and prayers. Many old time songs and melodics were rendered, and the evening performances were ended by singing "Good Nite, Ladies".

The men who operated the boats were called Captains. The Captains always took pride in mounting their boats. The dashboards of the boats were always decorated with flags and flowers along with the name of the boat.

The name of the first boat that went through the Canal was "Walk in the Water". The name of the pay boat was the "Katie Kellogg". This boat was very attractive, and very much desired. The "Katie Kellogg" left Port Delaware on the tenth day of each month during the boating season with twenty to thirty thousand dollars in gold and silver money. This was used to pay the plane and lock tenders and all the men working on the canal their month's pay. Never once was the "Katie Kellogg" robbed of a dollar. This showed the honesty of the people in those days.

All the water power of the Canal west of Lake Hopatcong found its way to the Delaware River. At Phillipsburg it operated a large sawmill on its last flow which brought the canal men and the river men very close together.

THE RAFTS ON THE DELAWARE

During the decade following 1880, the Delaware River was the center of the hemlock logging industry in this area. Log rafts made up of hemlock logs to be delivered to the sawmills to be sawed into lumber for building purposes were floated down the Delaware.

The rafts were composed of logs from twelve to sixteen feet in length and from twenty-four to thirty inches in diameter, rarely varying two inches from this diameter. The rafts were approximately one hundred and fifty feet long and fifty feet wide.

The logs were bound together by means of a birch stick about two and one-half inches in diameter, which reached about one-third of the distance across the raft, when the logs were side by side forming the raft.

Two holes, approximately three inches apart, were bored with a one-inch auger, and the birch stick was placed between holes, each of which was about three inches deep. Another birch stick about one inch in diameter, "U" shaped and resembling closely a staple, was driven over the former birch stick into the three-inch deep and one-inch in diameter holes in order to form a driving fit. By this means, the entire raft was bound together. Long logs and short logs were interjoined in order to make each of the cross bars of the raft the same length and to straighten the ends of the raft.

The birch stick about thirty feet long, tapering from six inches on the larger end to two inches on the smaller end provided the steering mechanism of the raft. The steering mechanism, or rudders as they were more commonly called,

The rafting of logs continued into the twentieth century. Trips were made during the spring and fall high water stages.

Just down stream on the Delaware River from the Andover Furnace was the entrance to Hagerty's saw mill. Note the close proximity to the canal.

were strategically placed, two being at the front end of the raft and two at the rear. The longer end of the birch stick was hewn flat, providing a space about four feet long and extending halfway across this space was bolted a hickory board six feet long, one foot wide, and one inch thick. Eight feet from the leading edge of this hickory rudder a hole was bored with a two-inch auger on the edge of the board. The completed rudder was attached to the raft by means of a series of four holes, eight inches deep and two inches in diameter, at the predetermined rudder positions which were ten feet from each of the four edges of the raft. Birch pins, two inches in diameter, were dropped into holes bored in the birch sticks which formed the handles of the rudders, and the handles, in turn, were inserted into the aforementioned holes which were drilled in the rafts. With these eighteen-foot tiller handles and the six-foot rudders, the raft was safely guarded by the river by the rangy, six-foot river men.

The river men, numbering four to a raft, were all practical, experienced hands, who know every shoal, rock, and bend in the river. In order to house their gear, consisting chiefly of their rubber coats, lanterns, and food for their journey, the men built a temporary shelter on the raft. Each man was provided with a coat and an ordinary coal oil lantern, since battery lamps were still unknown. By this meager light, the river men guided their rafts over the sometimes silent, sometimes roaring river.

The length of time it took them to make the cruise depended on the water level of the river. Both low and high water levels presented problems. During low water, care had to be taken to clear the shoals and the rocks, while in times of high water, it was necessary to use caution on negotiating the tricky bends which are so numerous on the Delaware. The average trip was completed in times varying from two days and one night to two days and two nights.

Extra logs were always rolled onto the raft and placed crosswise to provide a resting place for the skippers of the rafts. The river men enjoyed nothing more than to sit on one of these impromptu seats and puff happily on their old clay pipes, while nipping generously from the warming bottle which was always a standard part of their river equipment.

The port of entry in Easton was the Zearfoss-Hilliard Lumber Company, which was located on the Delaware near the Bushkill Creek, at the approximate location of the present Bushkill Street Bridge. The river men considered it as one of the most favorable spots at which to dock. Secure snubbing posts were provided and the rafts were tied up to these by means of stout hemp lines, while the logs were drawn out of the river and taken to the sawmill to be made into lumber.

The most dreaded areas of the entire Delaware were the bridges that crossed the river in the vicinity of Easton and Phillipsburg. The piers of both the covered bridges which formerly crossed the river at Northampton Street and the railroad bridges downstream a short distance, were constructed with sharp leading edges, in order to relieve the pressure of the water. In the spring of 1888, during a period in which the Delaware was at flood stage, one of the rafts struck a pier and was split into two parts. Some of the logs were salvaged and tied along the shore, but many logs floated free and were carried down the river by the strong current.

The next port of call farther down the river was D. W. Hagerty's sawmill, and, although other logs were carried as far south on the river as Frenchtown, Lambertville, Trenton and even into tide water, this was considered one of the last ports of the journey. The dock and Wint Hagerty's Sawmill, as it was called, lay along the cinder bank of the river, some two hundred yards below the Andover Furnace. Hagerty's dock proved to be a popular spot for swimming. A springboard was improvised. The tiller handle was lashed to the raft flush with its surface and with the rudder blade extending over the side in an inclined position.

An interesting sidelight of the journey down the river on the rafts were the Lamper Eels. The Delaware at the time was clear and clean, and the bottom of the river bed was easily visible. The eels, in an attempt to extract the hemlock sap from the logs, would inadvertently trap themselves by fastening their teeth in the wood. In order to remove them, it was necessary to strike them with a stick.

Hagerty's Sawmill was a popular stopover in both summer and winter. Here the Morris Canal emptied its overflow into the Delaware River.

The logs were removed from the river by means of a narrow gauge railway, which was laid from beside a pond of water near the sawmill on an inclined plane into the water. Enough clearance was allowed under the surface of the water to enable a log to be floated over the railway car, which was small and narrow gauge and fastened to it. The car with the log attached was then drawn from the water with a long cable which was pulled by a team of horses. The log was then carried by means of the car to the pond near the mill and was rolled into the pond.

This pond was known as the Log Pond, and it covered an area of about an acre of land. It could float the equivalent of approximately three rafts. The water for the pond was provided by means of the overflow from the Morris Canal, and the depth was great enough to float the heaviest log in the raft.

The sawmill was located at one end of the pond with easy access to a waterfall which was built there. The power to operate the mill came from an undershot turbine system of water which was driven by the force of water from the canal.

The logs were taken from the pond by another railway similar to that of the river bank, running from the bottom of the pond to the interior of the mill room. In this room, the log was placed on a frame which was the same length as the log and ran by the huge circular saw. With the log firmly clamped in place, the saw was set in motion and the initial cut was made on the side of the log. The outer cuts, taken from the four sides of the log, were called "slabs". When the initial slab cuts were made, the log was squared up on the carriage and the process of sawing it into lumber and scantling began. The waste lumber, such as the slabs, was sawed up into short lengths and sold as firewood at about one dollar per horseload. The sawdust, in turn, was sold to be used as packing in ice houses, bedding for horses and cattle, and for fertilizer in the gardens and fields.

Through the proximity of the canal and their interest in the milling of the logs, the raftsmen soon became acquainted with the boat Captains from the canal. Both types of men were titled as Masters of Inland Navigation.

Raftsmen often stayed overnight at Hagerty's Mill and enjoyed an evening of companionship with the boat Captains. The usual place for this joyful gathering was at the Change Bridge which was close to the mill. Parties were held in the moonlight, providing gaiety through the otherwise stiff and unyielding 80's.

These moonlight parties on the Change Bridge were composed of the boat Captains, their wives and families, and the river men. Music was furnished by the Captains with guitars, banjoes, violins, piccolos, and accordians. The only stage light provided was the moon. The party generally started the festivities by a rendition of "When The Silvery Moon Is Beaming", as more or less of a salute to their source of light. Entertainment consisted of the chording of many Southern and Irish melodies, as well as heel and toe dances, polkas, the Quadrille and waltzes. The party lasted about three hours generally, and the joyful evenings would be brought to a close by the singing of "As We Sat in the Evening By the Moonlight" or some such similar refrain. About two of these moonlit parties would take place each summer.

Gradually the hemlock forests of the upper Delaware petered out, and logging shifted to the Middle West where large hemlock trees and more abundant facilities were found. The rafts no longer came down the Delaware, the sawmill ceased its operations, and eventually in the early 1890's high water destroyed the last vestiges of that once colorful and happy period.

Though the physical symbols of the rafts of the Delaware were no longer evident, the memories of peace, happiness, and prosperity which were in evidence through the joyful '80's still lived in the hearts and minds of those who knew of logging and moonlit parties for many and many a year.

ALONG THE CANAL

The shipping of coal over this canal was a spring, summer and early fall occupation. Each year about 600 boats were engaged and about 1000 mules and 300 horses furnished the pulling power for these boats.

The people who were engaged in this industry were mostly white pioneers with the exception of about 5 percent colored people who had come North after the Civil War and considered it a matter of gratitude to help in the shipping of coal that their Emancipators might be protected from the cold biting winds of the winter.

After navigation ended in November of each year, all the white people retired to their homes for the winter, which were located in the small towns and country districts along the waterway. Some of the colored folks converted their boats into living quarters and remained in them for the winter.

The canal itself was all beauty. The towpath bank and the berm banks alike were lined with all kinds of trees which were mostly weeping willows and the plane house towers, which could always be seen in the distance, added to the waterway.

Sometimes during the summer heavy rains would cause the canal to overflow its banks and cause a washout, which would tie up navigation in both directions until the break was repaired.

The white people always mingled with each other, and after a man made a few trips over this waterway he was better known than if he had served a term in Congress. The colored people never made any attempt to mingle with the whites but were very clever in trying to imitate them, which made them very popular among the white folks.

This waterway with its many charms made an ideal place for love, romance and fun. As I remember one of these boats was operated by a Mr. Thomas Brice, commonly known on this waterway as Captain Tom. Captain Tom was a man rather tall, probably 5 feet 10 or 11 inches, slender built, a small mustache and his hair were equally divided between silvery gray and the dark color of years gone by. He weighed about 150 pounds. His face always carried a smile, and his eyes shone with such a gaze

Captains were proud people and usually dressed well. The towpath also served as a bicycle route.

The beer boat of Easton's Seitz Brewery plied the Delaware, Lehigh and Morris canals. Location: Mauch Chunk, PA.

as to have you under the impression that he took you all in a glance. He was very congenial, always pleasant and kind to everybody.

Sometimes it was hard to secure boys to drive the teams. So Captain Tom, in order to keep moving, had his daughter Una Brice, a girl of 15, make a few trips with him during the vacation season. Many other girls and boys in those days helped their fathers out during the vacation season. Una's mother, seldom making any trips, usually stayed at home and prepared the good things to eat, with which she supplied the boat each time it came through.

Una was a girl that always carried a very neat appearance. She had dark brown hair that was inclined to be wavy, dark blue eyes, and her cheeks and lips were well supplied with the scarlet that gave them the complexion that only Nature alone can give. Rouge and cosmetics in those days were unknown. She always had a smiling countenance and the tiny dimples in her cheeks were the attractive features that invited you to take another look. Her mode of dress was always impressively simple. All the boys and girls went bare-footed during the summer months, and Una was no exception to the custom. She wore a plaid gingham dress with a wide ribbon around the waist with the ends long enough to reach halfway down the skirt. The skirts in those days were worn long enough to reach halfway between the ankle and the knee. She wore a wide brimmed straw hat of a very cheap quality with the streamers hanging from the sides, which she drew down and tied under her chin, giving it a poke bonnet effect, and her long hair hanging almost to her waist was a very conspicuous and attractive feature.

On this Canal ran a beer boat. It served its wet goods at the locks and planes, and to the scow men along the wharves. Sometimes when the beer boat would tie up for the night along with the other boats, some of the boat men would indulge in the wet goods to excess.

There also ran a pay boat on the Canal, the "Katie Kellogg", and it was in command of Capt. I. P. Morris, of that old Morris Family who was responsible for the two million dollar corporation that bore their name, "The Morris Canal and Banking Company", which exists now in memory only.*

Capt. Morris was under the average height for a man. He had silver gray hair and side whiskers, and during the summer wore a linen suit and a blue cap. The "Katie Kellogg" was a brass mounted boat built on the order of a small steam yacht with compartments enough below deck to accommodate all aboard. It was painted white, striped with green letters and always carried the Stars and Stripes, waving at its bow. Two sorrel horses furnished the pulling power, driven by a colored boy with a darky at the tiller handle.

Arthur Bloomfield acted as secretary for Capt. Morris. He did all the paying and hauled all the money paid out for the Company. Arthur was a man of 20, probably 5 feet, 6 inches tall, and had straight shining black hair with a blush in his cheeks, brown eyes and a pleasant smile which made him a popular young man from Port Delaware to Jersey City basin. In fact, very popular when any wet goods were in evidence.

Another coaler that navigated this waterway was commanded by Capt. Wm. Hudson. His son, Farley Hudson, a boy of 16, drove the team. He was a good sized boy for his age with dark hair and grey eyes. Farley had been attending school every winter and continued his studies in the summer while helping his Dad on the canal. He seemed to be interested in mathematical problems of which he seemed to make it his hobby.

Amos Wardell also ran a boat for a livelihood. His wife spent most of the summer on the boat along with their daughter, Jenny, who put her time in along the towpath looking after the team. The feeding of the teams was done in a simple way—hanging a basket under the animals' noses, held by a strap served as a good meal for the team while they were engaged in their work.

Warren Kersey, a tall, light-haired man of very attractive appearance, and son of a wealthy brewer, was spending the summer at the lake. He became much interested in Una and

---

*The author can find no evidence that Capt. Morris' family was responsible for building or financing the Morris Canal.

sought an introduction to her. One day after a short acquaintance, while Una was feeding her pet lambs and swans, Warren endeavored to take advantage of her, which Una resented to the height of a struggle, which ended by the interference of a friend. Una afterwards learned from Jenny Wardell that he was a married man and tried to make love to every girl he met.

One fine day as Capt. Tom Brice was running the seven-mile level with a heavy load of anthracite, Una was beside the team to see that they did their best in order that they make a long run that day. Capt. Tom was at the tiller handle and was going along nicely when he saw the cabin window open. He immediately left the tiller handle and down he ran to close the window, but before he could return to the tiller handle the boat was aground. With the boat hard aground, the team strained every effort to draw it, but the heavy pull on the towline only resulted in the parting of the line. With the boat aground and the towline broken, navigation now was tied up in both directions.

Una now, for the first time, realized what an awful fix she was in. Looking ahead she saw a boat coming along lightly. It was drawn by a fast team and was making full speed ahead. As the team reached Una's place of misfortune, a young man stepped up from beside his team and inquired what the trouble was. Although Una knew the young man was Farley Hudson, she had never met him personally. After Una explained the cause of the mishap, Farley quickly secured the broken ends of the towline and with the skill of an able seaman spliced the line and was ready to hook up the team when he saw that Una could not control the team any longer. He ran to her assistance, and taking a rein strap in each hand, the team swayed around each other in such a manner that Una found herself in the arms of destiny. After Farley had gotten the team hooked up and Capt. Tom had pulled his boat off ground all was ready to resume navigation. When Una turned to Farley and said "Thank you, Farley. I hope some day I will be able to repay you for your kindness", and all was over for that day.

The colored folks were always sure to have their day. Tommy Duncan and his wife and family ran a boat. They had a boy named Sammy who usually drove the mules. Mammy Duncan always stayed on the boat and did the housework. She always wore a dolly vardum dress and a red bandana and would hang the clothes on lines stretched along the deck of the boat on washday.

Mahlon Jackson and his family too boated for a livelihood. Capt. Jack had a daughter named Monday. Monday and Sammy Jackson were pals and these two families were always the prime movers at the Camp Meetings. Monday always had a desire to feed her little pets. One quiet Sunday afternoon a group of half grown goats were gathered on the towpath. Monday after feeding a very gentle goat, tried to bring it on the boat. When half way up the gang plank the goat rebelled and butted Monday off the gangplank into the Canal. Monday's mother, seeing the incident, screamed and yelled which attracted Sammy, who jumped into the canal and saved Monday.

The first Sunday in July of each year was the date set for the annual Colored Folks Camp Meeting. It was held in the grove along the canal just outside of Washington, New Jersey. They held forenoon, afternoon and evening sessions. Their music, singing and dancing were always enjoyed by the white people. Al Simmerhorn always led the singing when they sang "Climbing Up Dem Golden Stairs" and after Monday appeared in the role of Topsy and put over some very nice steps in dancing, the entertainment usually ended with the colored folks singing "Good Night, Ladies".

The last Saturday of August was the date set for the annual Harvest Home to be held. This was attended by all the white people. Good eats, music, singing and dancing was the program for the day. Through a friend, Arthur Bloomfield met Miss Una at the Harvest Home. After a short acquaintance, one evening Arthur offered her a proposition that if she would consent to an engagement, he would some future day make her Queen of the "Katie Kellogg". Una did not seem overjoyed with the proposition although her friendship with Arthur was always cordial.

The summer passes and the winter brings on its social functions which were just as interesting as they were in the summer. As the years would come and go bringing joy and sorrow alike,

The **"Katie Kellogg"** *was well stocked with provisions. Usually the boat was pulled by horses rather than mules. Location: near Lake Hopatcong.*

Arthur suddenly was obliged to abandon his position on account of his extravagance at a critical time. This made it necessary for Capt. Morris to look for another secretary. Having seen some of Farley's work, he persuaded him to take the position, which Farley did, and served his duty with credit to himself.

The following winter the day was chosen when Una was to make good her promise to Farley, and on the evening after the ceremony was over Farley received word that Capt. Morris had died suddenly. This brought sadness to the young couple for a few days. The following week Farley received notice that he was to take Capt. Morris' place as Commander of the "Katie Kellogg" and the ruling spirit of the Morris Canal and Banking Company.

On the last week in June, Una, now Mrs. Farley Hudson, decided to make a trip on the "Katie Kellogg". At the locks and planes when she stood on the bow of the boat along with Farley, the people gathered to offer her their congratulations, presented her with large bouquets of roses and greeted her as Queen of the "Katie Kellogg".

While the Canal age today exists in memory only, it must be conceded that it laid the foundation of this great country. The trail blazers of the Canal age were responsible for the opening of vast deposits of natural resources and wealth that meant so much to humanity that it became the envy of the entire universe, and should never be forgotten by the generations to come. For such was the life of the people of the Morris Canal.

HARVEY MOWDER

Age 78 in 1976. Born at Port Colden in May 1898. Son of canalboat Captain George Mowder and grandson of Captain Daniel Mowder. Now living in Washington Township, Warren County.

Chapter Thirteen

Harvey Mowder

"Some folks call me 'The Sage of Warren County.' "

An interview with Mr. Harvey Mowder by Mr. James Lee—September 15, 1975:

What is your name?
My name is Harvey S. Mowder.

How old are you, Mr. Mowder?
Right now I am age 77.

What was your connection with the Morris Canal?
My father was a boat boy for his dad on the Morris Canal and he went to work on the canal when he was thirteen years old. He was born in 1875, went to work in 1888, and worked until 1898 on the canal with his father, Daniel Mowder. He was known as the owner of the blue mule—Dan Mowder's blue mule. It was used as a spare mule because Grandpa Mowder used stallions, grey stallions to pull the boat and "Old Blue" would ride the boat up on the front end and he would bring the empty boat home alone and the two stallions would ride coming home.

Did you have any other relation on the canal?
I had an Uncle Samuel, who was my father's brother who worked on the canal pretty much as my father did. On the other side, I had a great uncle whose name was Archibald Updyke. He ran a boat on the canal, too. He was from Brass Castle in the township of Washington.

Did you ever make a trip on the canal?
I made two trips with Great-Uncle Arch; one in 1908 and one in 1909. I was ten years and eleven years old. I made one trip to Newark and one up the little side canal where a tugboat pulled us. And then we floated back down on the outgoing tide and locked into the Morris Canal and on to the filling station in Phillipsburg, New Jersey.

Did you have to work your way there or did you ride along as a passenger?

I just rode along and fished from the rear end of the canal boat and had just a great time in general; walked awhile and rode awhile.

Do you remember what you might have ate on the trip; what your food might have been in the morning; what you had for breakfast?

Well, we had ham and bacon which had been home-cured and we had steak the first couple days out which had been purchased at one of the canal stores. I think we purchased steak one time at Rockport, Cappy Hill's Store at Rockport. We also had baked beans. We did a lot of cooking in a big iron kettle on top of a stove which was called a canal stove and faced upon the deck of your canal boat.

That's interesting to think that your grandfather, your father, and you also made trips on the canal. But you never worked on as a boat boy?

No. I never worked on it. I was a little bit too young, although my father was thirteen when he went to really work on the canal. I always have had rather a love for the canal. It seems funny, but one time I was struck by the Muses and I wrote what is considered as the Country Poem. It is a poem called "Warren County" and in it I referred to the Morris Canal. You see I was born in Port Colden and I just happened to insert that into it. I don't know how I wrote it; it just come to me. It was no effort. There is a certain section of it I would like to read:

> But I see again Main Street of Port
> Where the Green-leafed Maples meet
> And Starkers Lilacs bloomin' in rain
> Makes all those pathways sweet.
>
> Then there's the view of the Old Canal
> Windin' along the Hills;
> From the Lock on down to Dilt's Bridge
> Then on to Bowerstown Mill.

A lock prepared to receive an ascending boat. Location: Lock 6 West, Port Colden.

We boys are all in swimmin';
The water is warm and fine;
Then we all walk home together–
That Old Gang of Mine.

But "Death" has closed so many doors
That I don't feel I dare
Go back again to County Warren
With no one left to care.

Makes no difference where I wander,
No difference where I roam,
There's still that spot in Jersey
That to me is Home Sweet Home.
(And part of it is the canal)

Then when at last my Journey's O'er,
My sun sets in the West,
Please "carry" me back and cradle me
In those Hills of Warren I love the best.

And that is the end of my Warren County Poem. It has been sort of adopted by the County of Warren as their official poem. It is on files in Trenton and when Warren County had its 150th birthday celebration, I was part of the service and spoke in the County Courthouse as the sort of some official historian of the county without portfolio, as County historian. Some folks call me "The Sage of Warren County".

Well, I think you have earned your title.

I don't know, but that is what they call me just the same.

That's good—that reflects your love for the canal, too.

It's part of me. There isn't anyone who lived along, skated onto, swam in it, fished and paddled the canoe in it, and had trips on the canal boats. I even had a boat horn that my Grandpa had on the canal and you can still blow it and holler out for high planes. That is, if you are a canaler.

Were there any stories that your father told you that weren't in that story that you gave to me?

There is one that stands out in my mind. It happened up near Rockport in a thundershower in the afternoon. My dad was out holding the horses and going along and all of a sudden, the lightning struck in the canal proper! It was close to the horses and they were both struck down as well as my father. They were up again in a second or so, but severely shocked, but not permanently injured—or not even injured at all, just scared half to death. But the mules, or the horses, stood fast. Old Blue didn't know what to do. He was with my Grandpa there on the boat. He just laid his ears back and Heeee-Hawww. None of them were killed or hurt. Then I heard my father tell at one time there were some men or boys up on a bridge and they were loaded with rocks and they were going to rock this captain of the canalboat. When they came through, (course, my dad or my Grandpa didn't know anything about it), they got up there and saw the boys with the rocks. And they were huge rocks; if you had been hit on the head with one of them, it would have killed you. All of a sudden, one of the leaders of the group says, "Don't throw any rocks. For that is old Uncle Dan!" We don't know who they were looking for, but they just went on and minded their own business. Some of the old Paterson gangs must have had it in for one of the canalers.

Who used to bake the homemade bread?

Mrs. Lanning. Mrs. Melva Lanning over here at Bowerstown. She also baked bread for the boatmen, and S. W. Nunn's father, that was Alfred Nunn, Sr. That old building is still standing in Port Colden, and the Port Colden Consolidated School, the township school, is built on the old basin of the canal, at least part of it is, and that is where they used to build boats for use on the canal from the oak. My grandfather Skinner was one of the operators on the sawmill up near #6 Plane and they sawed the oak planks and floated them down to Port Colden to be fashioned into canal boats by Mr. Peer and a number of his boat builders.

They also had a boatyard in Washington, I understand.

There was a small boatyard in Washington, but I don't know who owned it. There was another small boatyard at Port Murray, but all the wood it seemed was made and cut for building the boats was sawed at the plane at Port Colden. That was a canal operated sawmill. The water force coming down the level of the canal, and that was #6 Plane, would operate the waterwheel that ran the saws. I heard my mother say that she used to take the lunch up to her father at noontime while he was operating the saws up there.

Did you ever hear your father mention any songs on the canal other than the church songs that his father taught him?

My father always mentioned church songs that Grandpa sang. His greatest favorite song was "When I can Read My Title Clear to Mansions in the Sky". That was sung at my Grandpa Mowder's funeral and also it was played quietly on the pipe organ at my father's funeral.

Mrs. Mowder: Don't you remember him telling about going to market down underneath a place somehow?

The Plane went down to the market, and underneath the market and the people used to throw out their decayed vege-

To the right is Newark's Centre Market, under which the canal was located.

tables and their liquids and so on, so that it would hit the boatmen when they went through. And there used to be quite a mess when they got through from underneath the market from the produce men that would throw the materials off there. They did that so as to get rid of them and too because then they would not have to pay to get the stuff carted away.

I understand that there were a lot of rats down there. I guess it would be because of the garbage down there.

My father would tell about the rats that were down there on the meadows of Newark. He said they were nearly the size of cats. They were ship-rats. They were not the common rodent rats. If you tied up your boat there, they used to even come up the rope that was tied to try to get on to the boat. The only way we could keep them away is to put tin around the boat which they could not get around, around the tin. There was a regular slip of tin around the boat.

Harvey, now we have taped this. You don't care if I use this for public dissemination in any way, shape or form, do you? And I would like to record your father's story in the Library. Do you have any objection to that?

None whatsoever.

I think it is a fitting story.

And proud too.

I think that people in time to come will enjoy reading that and hearing about your father and your grandfather.

Women added a domestic flavor to life aboard a canal boat. Flags, hanging from a tiller or dasher were not uncommon.

Mules drank directly from the canal, especially in widewaters where the bank slope is gentle.

GEORGE MOWDER

Born at Port Colden in 1875. Died in Washington Township, Warren County, in 1968. Began as a driver in 1886 at age eleven and eventually was a canalboat captain.

Chapter Fourteen

My Five Years on the Morris Canal

As told to Mr. James Lee by Mr. George Mowder
November 11, 1953

In the year of 1886, A.D., at the age of eleven years, I started my career in life as a boat boy.

Before telling anything about my experience of those five years, I believe I should tell what I can remember about the Morris Canal. It was a waterway across the northern part of New Jersey built to haul coal and other commodities across the state to tidewater; also to deliver coal to the many towns along its route.

It was one hundred and two miles long. The grade of the canal ascending from Phillipsburg to Port Morris, through the following towns to the best of my knowledge: Phillipsburg, Green's Bridge, Stewartsville near New Village, Washington, Port Colden, Port Murray, Rockport, Hackettstown, Waterloo, Stanhope, to Port Morris to Lake Hopatcong. Then it descended from Lake Hopatcong to Jersey City through the following towns and cities: Shipping Port, Wharton, Drakesville, Dover, Powerville, Rockaway, Boonton, Montville, Beavertown, Lincoln Park, Little Falls, Paterson, Bloomfield, Newark, also near Montclair and the Oranges, then to Jersey City.

There were 120 boats on the canal at that time. All the boats hauled coal at that time except a wood boat owned and operated by William Black, better known as "Billie", a flour and feed boat owned and run by William Thompson, the name of the boat was "Jim Bird", a beer boat which ran through to Dover, and also a lime boat that delivered lime to the farmers as far as Hackettstown.

I will now list the planes, locks and levels. A level was the channel of the canal between a lock or plane which lowered or raised the boat from a higher or to a lower level.

The lock and plane is somewhat difficult to explain without a picture, but I will try. The lock was a walled up structure a little longer and some inches wider than the boat. The boat comes in the lock when the lock is empty, then the gates are closed. Then the water is let in through a wicket and the lock is filled. The upper gate is then let down and the boat is on the higher level. When the boat is going downstream, it comes in when the lock is full and is lowered to the level below.

The plane was an incline of about twenty degree grade, approximately 500 yards more or less in length. There was a car that carried the boat up this hill to the higher level. The lift was, I think, perhaps 50 to 100 feet. The car ran on a track like a railroad, only ties or waybeams, as they were called, ran parallel with the incline. They had a wire cable attached to it. The cable was about 2½ inches thick, and this cable ran on pulleys about 12 feet apart, up and down the incline and then to the plane power house, which was operated by water power. It was a 12 foot turbine wheel propelled by a 50 foot drop flume about 5 feet in diameter; this was fed by a long open flume constructed of wood about 100 yards long, 6 feet wide and 4 feet deep.

This turbine operated a large iron drum about 10 feet in diameter which this large cable wound and unwound around. The machinery could be thrown in reverse by a large lever and in the power house which was built over the machinery of the power plane. The boat was run in this car and snubbed fast by the brakeman and he signaled the man in the power house and he turned on the water, and the car and the boat were taken up or down to the higher or lower level.

This is a list of the locks and planes and levels as you left Port Delaware with your loaded boat. The first were three locks at Green's Bridge, next a plane #10 West, then a mile or so level, then Plane #9, and this was a two car plane, next was Plane #8, some distance out from Stewartsville. Between Plane #9 and Plane #8 was a two mile level.

After leaving Plane #8, we had a three mile level to a lock called Gardners Lock, which raised the boat into the seven mile level. This level took you to Plane #7 at Washington, going over this plane put you in a three mile channel which took you to the Port Colden Lock, going through you were in the short ½ mile piece of the waterway which led you to Plane #6. In this short stretch of canal was the beautiful Port Colden Basin. It was 600 feet square and was a grand place for boating, swimming, fishing, and skating. There was also a basin in the canal at Washington, known as the Old Boatyard, but it was not quite as large as the one at Port Colden. When you passed over #6 plane your boat was in the picturesque three mile part of the canal which took you to #5 plane at Port Murray. Then you were in the eleven mile level, a fine part of the canal that flowed through the village of Rockport, in which there was also a large pond, and then on through the outskirts of Hackettstown. There was a good-sized basin of water there also. From there you passed on to the Guard Lock which passed on to the Guinea Hollow Dam. You went a mile up this dam to Waterloo Lock which raises you up in the Waterloo dam. You passed through about 100 yards of the dam and up over Waterloo Plane #4 into a one mile portion of the canal. I would like to state here that in the Waterloo Dam there was a very wonderful spring. How they found this was, that when they drew the water out of the dam and left nothing but the regular channel of the Musconetcong Creek. This spring was on one bank of the stream. The canal carpenters made a large wooden tub or cask without a bottom and placed it over the spring. It was ten feet

The plane tender lived at the bottom of Plane 6 West, Port Colden. To the right is the tailrace from the turbine.

Boats leaving Plane 1 West entered Lake Musconetcong. The towpath had a bridge in the middle to permit passage of water.

The rounded stern boats were a unique feature of the Morris Canal. Location: Plane 7 East, Boonton.

high and three feet in diameter. When the gates were closed and the dam filled, there was a fine spring right out in the pond just where the boat entered the plane car. We always got a pail of fresh water there. Going in the one mile level from Waterloo plane, you went through a very rugged country to Plane #3, which was the steepest plane of all. This plane put you in a short piece to Stanhope Plane #2 and then a short level to Stanhope Lock, which raised you into the big reservoir at Netcong. This was a reserve body of water to feed the canal in a dry season. There were many thousand acres of this body of water. Passing through this reservoir you entered Plane #1 and went over into the summit level at Port Morris, and the water was fed in this level by a short cut up to Lake Hopatcong. This part of the canal was two miles long and ended at Shipping Port Plane #1 East, and the current flowed down and east.

We then went to Plane #2. After passing Plane #3, we were at Drakesville and Jacksons Lock that lowered you into the famous three mile level. Plane #4 was followed closely by Lock #2, Burds Lock. This also was called Fresh Breads Lock because the man that tended this lock always had fresh bread on hand. It was always the same fine bread. He bought his flour by the ton and did a wonderful business.

About a mile and a half and you came to Plane #5 followed quickly by the five locks at Dover, number 3 to 7. Seven was a guard lock. From Dover you passed over a four mile level into and over Rockaway Plane #6. There was a large pond at the head of the plane. Also, I should have stated that there was a large and beautiful basin at Dover.

From Rockaway you cleared Locks #8, #9, and #10 before arriving at Powerville. Lock #12 at the foot of the dam, just after the guard lock, put you in a mile level to Plane #7 at Boonton. Lock #13, Cornmeal Millers Lock, followed.

Again a mile, then you came to the two Montville planes. They were very steep and long. The numbers of these planes were #8 and #9. We were then in a four mile level going to Beavertown and over Plane #10 into and almost straight canal two miles long. At the end of it was the lock known as Mainses. They had a very fine and well-kept stable there. It was at the head of the seventeen mile level, also the name of the town there was Lincoln Park. A mile down this level at Mountain View was the Pompton Feeder. It was a nice piece of waterway dug by the Canal Company up to Pompton Lakes.

The next town was Little Falls; then on through the city of Paterson and then to Bloomfield. At Bloomfield there was a plane #11, also a lock. We then went into the five mile level and into and through Newark. There was a stream of water that came down from Eagle Rock upon Orange Mountain. It was piped from the mountain and run under the canal just west of Newark and emptied into what is known now as Branch Brook Park. The pipe was broken on the top just after it came out from under the canal bank, and we always got a pail of this fine water at this point. At the foot of this five mile level we came to Mike Somers' Lock and then into a short stretch to the Newark Plane. This was #12 East. It was a two car plane on the canal. High Street in Newark passed over this plane track about halfway from foot to summit. You were then in the town level and passed under the city from Market Street to Mulberry Street. At that time the corner of Market and Broad was the busiest crossing in the United States.

Boating continued in cold weather until the canal froze. Location: Lock 17 East, Deep Lock, Mulberry Street, Newark.

At the foot of this level you passed through the deep lock. It was about 40 feet deep, but I am not sure if this is correct, it could have been more. At the foot of this lock there was a lock that let your boat out in the Passaic River if you had a load of coal that was for any of the towns or factories along the river. After coming out of the deep lock, you were in a two mile channel and if you did not go into the river, you proceeded on down this channel to a lock called Dunc Hendersons, and passing through that you were in a short level that took you to the river lock, which put you into the Passaic River. You crossed the river at that point. Your team going over the big drawbridge and towing or pulling the boat. After crossing the river you went downstream under the bridge and turned into the cut. This was a channel made through the meadows about one mile long, and this brought you to the Hackensack River, which is approximately 300 yards wide. The team went over the long bridge, which was also a drawbridge, and pulled the boat across the river going into a lock which passed you into the eight mile level on the way to Jersey City. At the end of this level you were locked out into the Jersey City Basin where your cargo was unloaded on the large docks. The coal was used by all kinds of steamboats for firing their boilers to make steam. Then you started back to your home port to get another load.

Now all loads of coal did not go to Jersey City, but to all of the towns and cities along the whole canal.

Now I will try to tell you a few of my experiences as a boat boy.

When I was told that I would have to make a full hand with my father on the boat, I thought it would be fine and I felt like a full grown man, and was very willing to do my part. I began to get ready for the big adventure.

The night before, Mother and I had a long talk. She told me a lot of things which have been my guide through life. Two things I will state she said, "I wish that you would never use tobacco." And then she said, "I want you to promise me that you will never drink any whiskey or beer". All through the 67 years I have kept that promise, and I am glad I did. (And I hope that all

DANIEL MOWDER

Born at Port Colden in 1830. Died at Port Colden in 1908. A canalboat captain and owner of the famous "Blue Mule."

the boys and girls who read this will do as I have done) It pays 100 percent and that is a good investment.

The next morning I was off on my first trip on the canal. One thing I would like to say, before I go on is, that as I lay in bed just thinking, I heard Mother say to Father, "Remember, Dan, he is only a child yet". Looking back to that five years so long ago, I know that my father never forgot. He was a wonderful father. Just to state three different incidences, when the nights were long, dark and stormy, and I was plodding along through the storm and mud, above the storm and the wind, I could hear my father sing with his fine baritone voice, "When I Can Read My Title Clear to Mansions in the Skies" and "Other Refuge Have I None". It was more beautiful than I can tell. Another time when we were blown out across the marshes in the Hackensack River, we were out there for two days and nights, and I was very much worried. He would say, "Now, Son, don't be afraid; Our Heavenly Father will take care of us".

And once when we were the last boat in a tow going up the Passaic River, there came up a very hard shower and heavy wind. My father and I stood on the hinge deck of the boat; it

Captain Dan Kinney with his driver at Lock 8 East, Denville. The short tiller indicates the boat is loaded.

was very rough and things looked very bad to me. Father took hold of my hand and I looked up at him. He smiled and I was not afraid. I am mentioning this because in afterlife in meeting the storms that came to me along life's pathways, many times I have taken that great hand of our Father in Heaven, and He had looked down and smiled, and I have had courage to go on and do my best.

Now the one thing that worried me most was what I would do when we came to a change bridge. I asked Father about it and he said "Don't think anything about it. The mules will know just what to do". Sure enough, they did and I had no trouble with change bridges after that. By the way, a change bridge was where the towpath changed over to the opposite side of the canal.

Our first trip was through to Jersey City. Everything was new but I got along real good. I was a little worked up when we came to Newark. I had to leave the boat at the head of the plane at Newark and meet it when it came out from under Mulberry Street. I don't know how far it was, but I will say it was quite some over a mile, and again Father said "Don't worry. The mules will get you through all-right". *Good ole mules!*

A change bridge allowed the mules to go from one side to the other without snagging on or tripping over the towline.

We arrived in Jersey City and it was a great sight for me to see so many things, boats of all kinds, and then looking across the Hudson River into New York City, with its thousands of lights at night. It was wonderful to me. Next came the unloading of the boat. They did it with a derrick, a large bucket attached to the end of a large rope all rigged up with pulleys. Four men in the boat shoveled the coal in this bucket and it was hoisted up out of the boat on great piles on the dock. My job was to hold the guy line. This was a thin rope fastened to the boat, and then stop it when it had swung out over the pile and in some way dumped it. I got $1.00 for that job. Then the next work was to take up the lining plank and clean up your boat and start on your way back from another load.

All trips were pretty much the same, but we would have some new experience every day and that made it really exciting, with a lot of fun thrown in. Our shortest trip was to Paterson, and to all towns east, and in the Passaic and Hackensack Rivers. The average load of coal for a boat was about 75 tons. We made about twenty trips a year, and I think the boatmen would average about $45.00 a load. My father would give me $1.00 a trip and some spending money. I would get $1.00 extra for tending the guy rope while unloading, and for cleaning up the boat after we discharged the cargo. I managed always to save the $1.00 that my father gave me, and then get me a twenty dollar gold piece at the end of the season. But somehow or other that would get gone by spring and I would have to start all over.

The first year I worked with Father, we drove two mules. The next winter the big black mule died. Then we bought a young grey horse. He proved to be a very good animal. The following winter the little sorrel mule broke one of its hind legs. We did everything to try to save her, but at last we had to bury the poor little mule. In the spring Father bought a fine grey from a group of western animals which a man brought in. He broke in fine and we had a great team. One thing I will tell about them, we always fed them from nose baskets, and they soon learned where their meals came from. When they thought it was time to eat, they would stop and look around toward the boat to see if their dinner was coming. Most always Father would throw their feed baskets out to me. Then they would turn around and come to meet me. We always took the bits out of their mouths and let them take their time while they were eating. After all these years, I can see those two horses coming to me for their corn and oats. There were many fine stables along the canal where you could put your teams in overnight. The charge was 25 cents and you got hay for two animals.

Mules relaxed while boats went through a lock. The driver watched the proceedings and was ready to help if needed.

My father was a good cook and we had plenty to eat. The way we worked it when we were moving was: Father would get the meal ready and eat it. Then I would get on the boat and eat, and wash the dishes. It usually worked out very good. Sunday mornings we usually had a nice mackerel for breakfast and some beef or pork for dinner. The best and most tasty meals I remember was salt pork and potatoes. I don't think anyone could make pork and potatoes taste as good as my father could.

There were many things that happened during those five years, but it was the routine from day to day. I can say this, that I scrubbed our cabin every Sunday and put the beds out to air in the sun. And if we were in a place where there was a church, I would go to Sunday School in the afternoon. I should say that I only went to school one winter after I started to work on the canal, but I tried to keep up with my studies. I have walked many miles with one arm through the harness of the team and a book of some kind in the other hand. I always tried to study a little after we had put up for the night and whenever I could find time from my work.

I think I traveled up and down the canal a hundred times—that would be about 10,000 miles. I walked most of the distance.

I think that is all I can say at present about the Morris Canal. There have been a lot of stories written about it, and I wanted to make this different. What I have written is all true—everything except about Powerville. I may not have put in the proper location. There may be a few other locations of towns locks and planes a little out of place (I am writing this from memory after 67 years have passed, and hope I have given a true report of the canal and of myself).

The winter was a time when major repairs were made to equipment. Location: Plane 10 West. Dave Merritt (left) and Peter Lenstrohm (right).

CHARLES SNYDER

Born at Brass Castle in 1885. Died at Milford in 1976. Worked as a section hand for five years.

Chapter Fifteen

Charles Snyder

"I used to have a custard pie pretty near everyday. My mother baked me a little custard pie, and maybe a ham sandwich. I'll tell you, we were poor at home and we had nine in the family, but we had lots to eat."

The following is a taped conversation of Mr. James Lee interviewing Mr. Charles Snyder—December 12, 1974:

How old are you, Mr. Snyder?
Eighty-nine.

Eighty-nine years old. And what was your connection with the Morris Canal?
Well, I first worked with my daddy for a little while when I was younger, but then when I got up to about 18, I worked two years on the scow. Worked on the scow—that was my connection.

What was your work on the scow?
Well, we did everything. We one time had to go up to #7 and change that coil, that there tape. Every so often that year you had to change that. Only had to change it once in my time. We changed it on #7.

You say #7. You mean #7 Plane?
#7 Plane. You see, they'd wear in them there grooves when they ride up and down them wheels. And they'd change it—take a loop and change it form one end to the other. Then that made it on a different run.

How long did it take you to change that wire rope?
Took a whole day.

Could you do it in a day?
Yeah. Boy, that was the hardest job I had on there.

When did they change them—in the wintertime or in the summertime?

They changed it in the fall. The boat had to wait there till we got it done.

And that took a whole day to change that?

Yeah.

And what did you do with the old cable?

The old cable. Oh, we didn't take the cable off. We changed the cable from one end to the other so it would run in a different course.

Oh, you reversed the cable?

Yeah, reversed the cable.

I see.

Didn't want to. It would wear and that would be in a different run, see. And change it, no we didn't.

You've never changed a cable?

I never changed a cable, but they did have to do it.

But you've never done that?

It broke one time, you know, half way up and the boat went down the canal.

When it would break, how long would they repair that cable?

They would get a new one—put a new one on.

They would generally put a new one on. That cable was fastened on to the trailer car, wasn't it?

Yes, it fastened on the trailer car and then it went up and down to the drum. It went around the drum and it run, you see, (that cable was the main thing to pull that boat up and down), and it would run it down and then they would reverse that thing up in the plane to run the drum backwards to wind it back up. That's the way they done there; up there in the whatcha-ma-callit. I don't know if there is any gears in that thing, but that lever was pulled from one side to the other to the water, to either run in either in another place to run it backwards. I'm not sure about that.

Well, they had a sliding clutch on there. The turbine always went in the same direction, but the drum could be reversed.

Yeah, it could reverse.

The drum on which the cable is wound is located to the right and below the flume. Location: Plane 5 West, Port Murray.

Did you ever work on the turbine?

No.

You never had to go down and do any work on that?

Never worked on that, we didn't–see, that wasn't ours. That was Port Colden. When we went and changed that cable, we had help from the Port Colden gang.

Oh, I see. What was your section?

Our section was from the #7 Plane to #8. That was our section –down to #8, but we had to help on #8. But we never worked on the drum or anything like that.

What other type of work then did you do on the scow? Did you ever do any carpenter work?

Well, we built bridges, if you call that carpenter work.

I guess you would.

We built their bridges.

How would you build bridges?

Well, we built the bridges–First of all, Art Unangsts and I would frame it at Port Delaware and then we would go down and get it and bring it up to Broadway and then we'd take it the next day and build the bridge.

How would you take it up to the place it was going?

The canal company had to maintain the bridges which crossed the canal, whether for the public's or a farmer's convenience. Location: Curries Wood, near Greenville.

Yes, it was about hunting season when we built the bridge.

But, now, you assembled it at Phillipsburg and took it to Washington.

Yeah.

Now how long did it take that boat to carry it that far? Did it take eight or nine hours?

That was all day, pretty near. We didn't get to Broadway till pretty near four or half-past four. That took pretty near all day because I know the fellow I had with me (won't mention his name) but he stopped at every hotel along the way. He wouldn't steer it. And I never steered a boat in a lock before and this was part of the way up. And I said to Frank, the boss down there, "Would you give me a man to tell me what I am doing right getting that?" I said, "I've never steered it and if I bump her, I knock that lumber in the canal". He says "yes" and he gave me Bill Gross. Bill Gross went with me and we went in all-right; went in every plane and never touched; went in all the way myself. He was loaded, he was. He quit, too, through it.

On the scow.

On the scow?

Oh, we had a big load pulled up. There was a lot of lumber on one of those bridges. All the pipes and everything was in the bottom.

Did they have a carpenter shop in Phillipsburg?

I don't know. They had a shop that was where all this lumber was and all that.

And that's where you framed for the bridge?

That's where we framed the bridge—at Port Delaware—and then we'd go there and bring it and unload it and the next day we'd put it up. The next day after we'd get it, we'd come to Broadway with it, and then we'd take it up the next day and unload it. And then that was a day's work.

What bridges do you remember helping build or assemble?

Well, I helped build the one to Washington, off that way across the canal there; helped build that one, and I helped build the one that goes up to New Village to Montana and I helped build the main bridge across the canal on the turnpike there at New Village.

That used to be Route 24 at one time.

Yeah, that used to be a good bridge.

How long would it take you to take the bridge to Washington—a whole day?

No.

It wouldn't take that long?

To transport it, when you got done, you was done. When you unloaded, you was done. Because I know old Aaron came to me and he had to unload it and it was about 11 o'clock and he says to me, "Now when we get done, don't bother with Joe". He says he was the boss and Joe and Aaron Vough was the Supervisor and he says, "Jump off and go hunting. Your day's work is done." And he told Frank Richie the same way.

Well, that must have been hunting season then, or didn't they have hunting season then?

Now Mr. Mutchler lived in Broadway and he was your section boss?

He was the section boss.

Mr. Unangst–where did he live?

He lived in Low's Hollow.

In Low's Hollow–near what town was that?

That was near Stewartsville, a little bigger than what it used to be 'cause they are building houses all the time up in that way.

What other job would you have on the scow besides building bridges?

Well, we had, now, come a break or anything–I know one time I wasn't on the scow then, but was walking the towpath, the Morris Canal–the whole seven miles went out. I'll tell you where it was–right here below in-between New Village and Broadway. It went out and they claim that when the Edison broke out in the big stream of water; why now they claim that's where the canal went. They chased it and chased it and couldn't find where in the world it went. The whole bottom went out of the Morris Canal. Went right out and the water went out too.

Captain Bill Gross and his wife stand at the stern. Fellow workers also were glad to have their picture taken. Location: east of Green's Bridge, Phillipsburg.

And for seven miles there was no water in the canal then?

Well, they did get some in because they went right away as fast as they could go up to Broadway and put the stopgates in.

I see.

Like they do in the wintertime.

Stopgates in Broadway and where else did they have them?

Had stopgates in Brass Castle, up there.

Brass Castle. How about any in the New Village area?

No, none around New Village that I know of. They had them at Broadway and that held them both ways, you see.

Yes. What other type of work then did you do?

We loaded the scow with dirt and like that and cut grass and brush.

Where would the dirt come from?

Most generally, New Village would fill up with sand and things there in that creek and we'd have to wheel that on the scow and take that and dump it over the bank.

Oh, the creek would come into the canal?

Yeah, and that would fill up where we'd have to now take that out of there.

Would you drain the canal to do that?

No. we'd drain it in the spring. We'd wheel it out too. There was quite a lot of it in there too. No, we didn't drain the canal. They had a way, I forget how it was, but anyway, we'd catch it there and we'd wheel that out of there and take it and put it on the scow and then take it to a place and dump it on the bank.

Well, they probably had some kind of trap there in which that silt and sand would go in and they would have to keep emptying that out to make room for more.

Yeah.

Then it wouldn't fill up the canal channel.

Yeah, that was how it was.

I see.

Every fall we would go down there to Port Colden and our gang would go down there and clean that out all the way up. We'd be working there for a week maybe or for a couple of weeks. We'd be working there wheeling that out and up on the bank. Now, there's where we got some of it too. Had to take it away so you'd have a place to wheel it to the next time.

And you said you also cut brush too?

Yes, cut brush. We cut the brush on the bank of the towpath.

What were your hours on the canal that you can think of? When did you start?

Went to work at 7 o'clock and quit work around 5. Around eight to ten hours, it was.

Well, now suppose in this one day you fixed the bridge in Washington; you went from Phillipsburg to Washington with the bridge in one day; now how did you get home? Or didn't you go home?

Oh, we rode the scow. See, I went past my house when the scow would go to Washington. I'd go down in the morning to report and get on the scow and then I'd go past my place, but when it went down at night I'd get off. I didn't go down with them because they'd go down and tie the scow up and get the mules.

Where did you report at? Is Phillipsburg where you had to report?

No—right there on the scow.

Yeah, but where was the scow?

Right at Broadway. We tied the scow up every night right there by the bridge at Broadway.

Then how would you get home?

I had to walk home.

How many miles was that?

Well, how many miles I don't know. It's quite a ways.

Five, anyhow?

Yes, five miles.

And then in the morning you had to walk five more to get to work?

Yeah.

Then do a day's work?
Yeah.

Your hours in a day were generally . . .
I believe 8-10 hours, I don't know. Now let me see—7 o'clock to 5 o'clock, well, say 5 o'clock. We most generally ran in a little earlier. No, Saturdays we was supposed to quit, maybe we'd quit at 3 o'clock.

And you'd take your lunch hour too?
Yes.

Did you have a half an hour for lunch?
Oh, you had a whole hour.

Oh, a whole hour?
Yeah, a whole hour.

And you could do whatever you wanted to in that hour?
Maybe we'd go out and look for chestnuts or something.

Or berries or something, depending what was in season? Well, then it wasn't really too hard of a job, was it?
It was regular labor, but it wasn't too hard. No.

Work scows were placed at key locations along the canal. The work crews usually were able to return home each evening. Location: Dover.

When would you be laid off? Would you be laid off in the wintertime?

Oh yeah, there was no job in the winter. Laid off in the winter and start up again in the spring when they drawed the canal off—somewhere in March. In March, I think, or towards the first of April.

That's when they started the canal?

Yeah, that's when they started the canal. They would draw all the water out of the canal and what we had to fix the locks, we did. When we worked with Art, he said to me, "Go up to the house and get some cement". I got the cement and he throwed it. There was a dam there where it didn't run all out and he throwed it in there. Holy Moses, we got more fish—made 'em sick—floating on top of the water. He was a great guy, Art was.

Would Art work all year round then?

No. He was laid off too. In fact now, he might go if he had quite a few bridges to frame or something like that. You see, he didn't just frame our bridges in our sections, he framed bridges in other sections. But you know, there was a lot of bridges on the canal that belonged to the farmer. The farmer didn't know what he was up against. He would take so much and keep the bridge himself and there is where he was wrong. They had to keep them up and when they went down they had to rebuild them and do everything. It was quite a lot. There was a lot of them and they took so much money to keep the bridges in repair themselves. But, as I say, we had to go down and get that bridge and take it up to Broadway and then the next day take it where it was wanted and then you was done. Your day's work was done whenever you got it done, but I'll tell you—you worked! You build one of those bridges, you don't get done with your day's work too early.

How long did your father work on the canal?

Oh, now he didn't work too long. He walked the towpath for awhile, but he was a boater. He always was a boater—as a young man he was a boater. He boated from Port Jervis down to Newark; then be boated from Port Jervis down to

Trenton, down that way; then he run one of them freight boats. He liked that.

What kind of freight would be hauled on a freight boat?

Oh, store goods.

Store goods for stores along the way?

I didn't see it; I don't Know. That was before my time.

Your father was telling you about it?

Yeah, my father told me about it. That's where he told the names. I told you about that. Pope Rush went along and he went along and they hauled these all kinds of groceries. At Port Colden, oh man, they unload a lot there.

Where did they get their freight?

Down near Newark or somewhere along there. They would deliver to all the stores along the canal. It was just like a railroad. When the Morris-Essex Railroad came through, why, the canal was done. But they had the Porter Ale boat on there. I guess they'd deliver liquor and things with it. And beer and such. That was a dandy boat. Paul says that was a

"The Maid of the Mist," *also known as the beer boat or porter-ale boat, may have been a converted packet boat. Location: Walnutport, PA.*

rotten shame that we didn't keep that there boat because it was just like the Porter Ale boat, only it wasn't quite as big. It was built just like it.

That is too bad. I wonder what ever happened to it. You don't suppose it's there in the attic yet?

No. we don't think so. We think the Rush kids got it and broke it up. That's what we begin to think. We left it there and it was the biggest antique that ever was. You couldn't tell it from the Porter Ale boat. Only the Porter Ale boat was a good deal bigger.

Your father—what was his name?

Emmanuel Snyder. He told me there running that freight boat was really good.

Hard work, but it was a good job.

Yeah. They unloaded at every store, pretty ncar. You see there was a store at Port Colden and a store below the Plane and then Joe Adams' store at Brass Castle. That still stands there yet; the Grange has their lodge there. They went to Joe Adams' and then on to Ben Keck's Half-Way House. They called it the Half-Way House because it was half-way in-between the lock and the plane.

And where was that located?

In Broadway, at Ben Keck's house.

What about Stone's store? Do you remember Stone's store? What was his name, H.H. Stone?

Oh yeah, it was Hank Stone, wasn't it? I know it was a Stone and I know he kept that store right near there where they came down the bridge and the team changed towpaths. Now he run that store there and most the folks would stop there and buy up and then go up to old Ben Keck's. They'd get off there too. If they wanted to stay overnight, Ben Keck had a barn big enough to stable mules. That was what was below the plane. That was a store and a stable for their mules.

Well, Mr. Stone had coal there too, didn't he?

Yes, he had coal there and Mike Dowling did too.

That was on top of the plane?

Yes, that was on top of the hill along the turnpike. It was right along the canal though.

Did he have other things in the store? Did he have hardware and paint and things?

I don't remember, but I think he did. Seems Mike had everything. Mr. Stone had everything too.

I hear that Stones had good tub butter. That was one of the big things that the boatmen would buy.

I didn't buy nothing off him. After I moved to Stewartsville, he wasn't there any more. I lived in Stewartsville for 38 years. Whatcha-callit had it then.

Well, he died in 1911, I believe.

I moved there in 1918, when World War II stopped, and he was gone. There was other Stones there, Man Alive, it was full of them.

Was there much fighting in the canal that you know of?

Well, not too much that I know of. Old Poppy Pearson was a great scrapper.

How about Art Unangst?

Oh, Art Unangst was a fighter. He fought down around Green's Bridge with some guy, a prize-fighter—bare fists though. He was a scrapper, he was.

Didn't he go to Alpha one time after a lad?

Oh, yes. He was a good scrapper. I'll tell you—all the Unangsts were—even his sons were. Art Unangst was a good mechanic. He worked on the canal and he might have got paid through the wintertime, I don't know. They thought a lot of him.

How many men worked on the scow at one time?

Only two to a time.

Two and a boss—that would be three altogether.

Three altogether.

How many mules did you have pulling?

Two.

Did you always have two? Did you ever have one?

Always used two.

Did you ever use horses?

No, always used mules. I don't think there was a horse on the scow. Port Colden had two mules, two gray ones. Frank Piatt had two mules. In the fall, we'd try to get the boats off the canal before it would freeze up. You had to take your team and put them to the boats. Then Port Colden team went to it, the #8 went to it, but #8 only went to the top of the plane. We went only to the top of the plane. We did nothing below. We might have helped him out, change the rope or something. But now we put the mules to it as high as Port Colden and then there would be their teams. You would have 8-10 mules to it.

Well, the ice would form in the fall and sometimes the boats wouldn't be able to get out too easy so you'd have to break the ice. How did you do that?

Well, we would have a thing right on the scow in the front. I can't remember how it was but I think it was a sharp thing and it was up under the scow. It come up and hit on top, you know what I mean?

It didn't float or anything?

No, it was fastened right on the scow and the scow run up against it or on it. I don't know how. I think it was something sharp there and it cut that ice. It cut it pretty thick too. But if you could, as I say before, you try to get the teams on and get the boats through without using that. When you broke the ice, it followed them on up. Old Pop Smith was always in the last boat, always trying to get one more load. But we had to get them off as far as #6. Every section had to get them off that section. We went before Port Colden. The Port Colden team always helped. They helped in the fall of the year when we cleaned that creek out. They would help too.

How thick would the ice be that you could cut?

Sometimes it was pretty darn thick.

Couple of inches, maybe?

Well, not hardly. An inch thick we could cut all-right. But they loaded that stuff. It all had weight. I don't know if it was dirt or stones, but I think it was gravel. Somewhere near New Village we would fill it right up—in the back there. You see, the scow had a cabin in and a deep basement that you filled right up. You had to haul dirt in there sometimes. We wheeled dirt on that scow and when you got lumber, that's where you put the lumber. You'd pile that lumber way up. That's the reason why I was afraid that I couldn't steer it through the locks. I never steered a boat in my life.

You got good experience that day then?

Oh, I had an awful experience! I had Green's Lock, #10 Plane, #9 Plane, #8 Plane, and the New Village lock and plane. Holy Mackerel, I had good experience.

When you would take these boats that were stuck in the ice when you used the ice breaker and went all the way to Port Colden, that made a long day for you. You must have got overtime that day.

I don't think so. I think they went all the way to Port Colden because I don't think that they had an ice breaker up there.

By the time you got back to Broadway it was getting pretty late.

Didn't go to Broadway—never did. Whenever I went past my place, I always got off.

Where did you live?

Brass Castle, right below the store there, right down here below the red school house.

I thought you were living in Stewartsville.

No, I lived in Stewartsville after I was married, but then I didn't work on the scow. I worked at the Edison then. When I worked on the scow, I wasn't 21. I remember, John Mutchler was kind of a tough boss and when I was done, he said to me "Next year, you'll work next year?" (I was 21 years old that fall) I said to Mutchler, "I don't know. I won't work for you. I'll pick bones and buy rags before I'll work for you. Two or three times I caught you sneaking over the bank watching us and what we was doing when we was cutting brush. We was working hard, and two or three times you'd sneak around like that. I don't like a boss like that. Because when I work—I work!" And that was what he done. I never liked him for a boss.

Why did you work till you were 21? Why didn't you quit when you were 20?

Well, my father held me until I was 21 and took my wages, right up till I was 21. In them days you wasn't of age and you didn't get no money till you was 21. I think that was the best thing that ever happened because today the young kids get an allowance when they are in the cradle.

Instead of the son giving the father, the father gives the son money now. But in your day, the son gave the father.

I gave everything I made. If I went out and made a quarter, I handed it over. But when I went out to a festival of something, they'd give me something to spend. When I was 21, I could do as I pleased. Then I quit and a fellow gave me a recommendation down to Phillipsburg on the Railroad, the ones that went out to repair on the wrecks. We had to be called out at night and I thought I could go from home. The fellow was right here on the Central Railroad out here on South Main Street. He said, "Your recommendation is good and I'll hire you. I'll tell you where to go to get board." I was going up the hill and I says to the other fellow, "I don't want no board. I'll live with Pop and Mom. I'll get a job." I came back and told him and I'll bet that man jumped two foot right in the air and swore and says, "Your recommendation ain't worth nothing." And I says, "Okay." And I went home. I stayed with my father and mother right up till I was 32 years old–then I got married. I got a job in the Ordinance. When I got home there was this job, but I only worked two days and was laid off the rest of the week, so I went to the Edison and got a job there.

What was the average wages you got paid on the canal?

$1.20 a day.

How often did you get paid?

Why, I forget. Either every month or every week.

Probably once a month.

I think it was once a month. There was a pay boat down the

canal and it'd give checks out to Mike Mutchler and like that.

And you worked six days a week?

Yeah.

And there was no deductions or that either. When they said $1.20, you got $1.20.

No, no income tax. When I worked at the Edison, they paid you right out in money—no reductions.

How about shovels and picks? Did the company furnish those?

Oh, yes. They furnished everything. Brush cutter and everything.

How about your lunch? What would you eat for your lunch?

Well, just ordinary. I used to have a custard pie pretty near everyday. My mother baked me a little custard pie; and maybe a ham sandwich. I'll tell you, we were poor at home and we had nine in the family, but we had lots to eat. We had a place we raised potatoes and stuff like that. We always had two or three hogs, and my father would butcher for farmers and kill a beef for them. Then we'd get a quarter of beef off them. We had lots to eat. My mother says there is one thing—we can raise our eats and another thing is—there is lots of water I can eat between.

A page from the Monthly Time Book of the Phillipsburg boatyard, September 1887 to March 1891. The number of employees varied from month to month.

TIME. For the Month of July 1888

NAMES.	Total Time.	Rate per Day.	AMOUNT. $	Cts.
W M Carhart			65	
Wm Loderick			52	50
Chas A Butler	31	150	46	50
C H Miller	25	185	46	25
Daniel Stametz	25	175	43	75
Arthur Unangst	2	200	4	
John Allen	25	180	45	
Daniel Osborne	25	160	40	
Wm Wilhelm	25	"	56	80
James Price	25	200	50	
Alonzo Smith	25	160	40	
F H Elyea	25	130	32	50
			502	30

And there were nine in your family. How many brothers and sisters?

Four brothers and five sisters, but one died young.

What in your mind stands out the most about the canal? Any experience you had that was different from anything else? Were you ever scared?

I fell in the canal once. I was scared several times that way. I know a boy that fell in the canal and couldn't swim. I didn't know what to do. I couldn't swim either. But, by golly, he held himself up and I stuck a pole to him, he got it, and I pulled him out. We used to skate an awful lot on the canal and used to cut the ice. In the wintertime there is where we made a little money—cutting ice for the farmers. They'd hire us to saw and cut it and then you'd have to be careful skating or you'd fall in.

Did they ever play any tricks on you on the canal? Did they ever hide your lunch?

No. I had a dog that used to go with me—called him "Brandy". He was a dandy. I set my dinner pail down and he'd lay down right next to it. John said he could get that dinner pail and I said, "No, you cannot. I'll bet you on it." He couldn't get it either, unless he killed the dog. Boy, that old dog showed his teeth and come right for him. I could take my horse and wagon up to Washington and put it in the shed and that dog would lay right down by the feet of that horse. Then nobody could take that horse out of there. Frank Bend, who run the Livery Stable said, "Will you please put your horse down here? I can't move it. That dog will go at me." That's the kind of dog he was. He was given to me off the Morris Canal. A boatman gave him to me.

When you were working on the canal with the scow and a boat came along, what would be the procedure?

He'd either stop his team and we'd run over his towline if we was moving, or if we was standing still, we'd get ahold of the towline and take it over. I'll tell you, you had to be careful there or you'd get throwed in the canal too.

Suppose he had an empty boat. Then he could run over that towline quite easily? But, if the boat was loaded, he had a good chance to snag it.

Yeah. We very seldom ran over a loaded boat; they'd run over us. See, if we was coming down the canal and a loaded boat was coming up, we'd run a path on the other side and let them run over our towline. We'd go over on the berm side. Empty boats go to the berm side and their team was stuck and we'd run over it when we was going up. Any boat going up had to run over the empty boat's towline and stop the team and run over them because it would drop in the canal. The scow would stand still and then you had to pick up the towline to get across. The team went along with you.

What year did you start on the canal?

That was 1903 or so, until 1906.

Were there many boats working the canal at that time?

Oh, at that time when I was working the scow there were quite a lot of coal boats, a lot of boats going up and down the canal.

The scow allowed loaded boats to pass over its towline, the opposite of normal passing proceedings. The scow usually stayed along the berm. Location: near Hackettstown.

How about lime boats?

No, no boats carried lime. My daddy boated cordwood for a man, but they had to rent the boat. That wasn't run by the company. He'd take a load of wood down wherever they got this pig iron down here to the Durham Furnace. He'd take a load of pig iron back and a load of cordwood down to unload it at the Durham Furnace. That was when he was a young man, when he boated a lot. He run a freight boat then, him and Poke Rush. They had funny names. They were half way to Jersey City, and had to tell them the name of the boat. Poke Rush's was *"I'll tell you after awhile"* and Pop's was *"What's it to you"*. Of course, they called them crazy. But they come to find out, that was the name of their boats.

I imagine by this time the toll collector was about ready to call it quits.

Oh, he thought they were crazy, but it was the name of the boat.

Do you remember any toys or games that were used on the canal? They didn't have too much money, so maybe the parents made some of them.

All I know is, we used to play "shinny" on the canal with tin cans like hockey on ice. I don't know of any toys. They were always afraid the children would drown, so kept them away as much as they could.

Did your father ever make anything for you to play with?

Oh yes. He used to take elderberry and make whistles out of them. Chestnuts, too. Boy, oh boy, how they'd whistle. Another great thing he used to do was get birch and scrape it and chew it. The birch bark–that was great. One time I took the boys for a hike, I stopped and I saw this birch. I got my knife out and I scraped it and they thought that it was great.

Did you ever play with buttons on a string?

Yeah, we used to make them. You take a string and put a button on it and pull it up against a thing and it would whistle. We'd take two chestnuts and put them on a string and make one go one way and one go another. He used to make all those little games. He taught us how to play jacks.

He'd take a shave off a stick for us and lay them down on the ground and hit it with another stick and bat it in the air. The girls played with their dolls a lot. But, of course, I started to work when I was pretty young–9 years old. I went out of school and went to work for woman on the farm. I worked in the garden and in the yard. Belle Bullman was awful good. She had two girls and she put them to sleep in the afternoon, but she would get them up about 3 o'clock and say, "Now quit and go play croquet." So I'd go out and play croquet till supper, then we'd eat supper and I'd go home.

You said you had a little cabin on the scow. What was inside that cabin?

There was a stove down in there. In the fall of the year, you could go down there and eat your dinner and sit in the warm. There was this man who liked to play a joke on you, but if you got one on him then he'd get mad. I know this fellow on the scow played a joke on him. He filled the coat sleeve of his raincoat with water and tied it shut. When he went to go into it, there was water in it and the arm was half full, and he couldn't tell what it was.

Who did that?

A guy working on the mud digger. Old Bill Gross and the cook were quite proud of him.

Did Mutchler think that maybe you might have done it?

No, he knew who done it and he got mad. But he played jokes on him, but he didn't get mad. He said his mule could beat my horse. I had two men in my wagon one night and I said, "Come on–get your mules out!" and I beat him. He got mad. I beat his old mules.

Did they have a little table in that cabin?

No table. They had a place to lay down if you wanted. They had a blanket on it. Only one person could lay down on that. That was for him. He laid down there a lot while we'd be working. He had good men. He had no right to be sneaking down the back. He had no right to do it, but he done it–to everybody. Frank Richie was a good worker; so was Frank

McSickle, I worked with both of them. They was good workers whether he was there or he wasn't.

Did he ever tell you "Watch out. Don't let Aaron catch you?"

Oh, yes. He always said that. And Aaron didn't care. Why, Aaron Vough used to pick me up when I walked the towpath, in his buggy. Aaron Vough was awful good. I seen him after the canal shut up a couple of times. I talked to him. He said, "You never went back" and I said, "No, I wouldn't go back that next year because of the way it was."

Well, I never heard anybody speak bad of Aaron Vough.

Aaron Vough was a good fellow. I liked Aaron Vough.

How about songs or ditties? Did you ever do any singing on the canal?

No. I don't remember anyone singing on the scow. There wasn't any singers.

Maybe if you'd had a different boss, you would have done more singing. But you never heard any poems except—

Aaron Vough (third from the left) was a canal supervisor with headquarters at Phillipsburg.

You Rusty Canaler
You'll never get rich. . .

Oh! Everybody heard that!

You Rusty Canaler
You Son of a Bitch
Why do you follow
That dirty old ditch

That there we used to sing. They were pretty tough on that boat. Some of the boats were pretty good. But old Poppy Pearson was pretty tough. And Pap Smith, he was tough to the point he couldn't keep a hand. He told a white boy (not a nigger) who was out on the towpath driving the team, "You cook two eggs for you and two for me, and one for that dumb nigger out along the towpath."

What was different about Pap Smith?

His leg was off. That man was as swift as a cat. He'd jump off that crock and hit the tree and then he'd swing around whole and land on the bow; the bow would go out and the stern would come in and he'd get on it. He'd hit those mules a couple of cracks, and jump on the boat again. He'd do that right along. He was a great boatman. He had an awful good team, but he had one that kicked.

Did you know Peter Lenstrohm or Dave Johnson?

One of them Johnson's was the cook on the mud-digger. It seems to me, there was a young fellow who worked with Bill Gross on the mud-digger. The scow was most generally always with the mud-digger. I know Bill Gross—worked for Frank Piatt. And I got a book home with all that, Frank Piatt's gang and the scow, Easton-Phillipsburg pictures. But I don't remember much else. We worked every day. It was either wheeling dirt on the scow moving it or in the spring of the year, we'd swing the mud out of the canal onto the towpath. I was only 16 years old when I done that for Pop.

Well, if you think of anything and when I come over to see you again, you can tell me. Even if we don't tape it, you can tell me.

Yes. Okay.

WILLIAM "BILLY BLACK" McCULLOUGH

Born near Lake Hopatcong in 1834. Died near Lake Hopatcong in 1903. A canalboat captain.

Chapter Sixteen

Billy Black: The Converted Canalman

Billy Black has a boat on the Morris canal. He is well spoken of by all the boatmen. Long before I met him, I heard his name mentioned as a true Christian. "Billy is all-right," said an old canaler.

According to his own testimony, he was wild by nature. "I was of a wild nature" is a favorite expression of his. "I was the wildest kind of an Irishman." His wild nature drove him from his home at the age of sixteen. He says, "My uncle was good; my aunt was good. I had no need to go, but I did not wish to be controlled."

"I was a wayward child, I would not be controlled."

I would not be controlled."

From sixteen to twenty-six he was addicted to strong drink. About that time he gave it up. He was still a sinner. Sabbath-breaking, swearing, chewing tobacco, ("slobbering 'round" as he calls it) were his special sins. He lived in what he calls "a wild, rough, uncultivated state." He says, "I was not a robber, nor a murderer; but I was full of dirt and filth and foolish talk." Like other canalmen, he could talk for an hour about a mule, but he says, "When the Lord saved me, He put me to talk about other things—the things of His Kingdom." He says, "The less you have

---

*Author's Note: William "Billy" McCullough was born in 1834 of Scotch-Irish Parents in Scotland. He emigrated to this country when he was quite young with an uncle and aunt by the name of Black. It is not known if Billy was adopted by the Blacks or if he just took their name, however he always went by the name of Black. Billy and his wife had sixteen children, four of which died at child birth.

A group of boats from the Morris Canal await their turns at the coal loading pens at Mauch Chunk, PA. After 1871 most boats loaded at Port Delaware.

to do with the animal kingdom, the sweeter will be your spirit." When I reminded him that the birds were beautiful and their songs so sweet, he replied: "There are songs that give me more delight. They are the songs of the saints."

Even in his wild days Billy felt the strivings of the Spirit of God. He said: "I used to talk to the Lord, especially when in trouble. I asked the Lord to deliver me from the appetite of strong drink and He did. One day in Mauch Chunk, the Spirit of God strove mightily with me. I found out afterwards that my wife and other Christians were praying for me that day at my home in Stanhope, N.J. One morning at 3 o'clock I went out to feed my mules. I felt a good Spirit there. My whole body, spirit and soul were filled with a sweet peace and happiness. I will never forget it. It will go down to the grave with me. Some might have been deceived, and thought they were converted. It was not conversion, for I did not give up my swearing. I was not changed from my foolish talk and vulgarity and swearing."

There is no greater evidence of conversion than the dropping of the old habits.

Strange to say, Billy gave up drink and tobacco before his conversion. He says, "In 1861 I enlisted in the war, and said I would not drink till I came back, and I did not. I was living at a hotel at the time. The inn-keeper's wife said to me, "Don't you want your bitters this morning?" I replied, "No." I kept my word till I got through the army. Then I had an inclination to drink. My wife noticed it, and said, "William, you were a sober man when I married you, and if you take to drink, I won't live with you." I thought much of her, and would rather part with the drink than with her. A woman has a wonderful influence if she uses it for good, and she has a wonderful influence if she uses it for bad. A woman can do more with a man than a man can with a woman."

Billy Black's wife strongly influenced his personal life. She encouraged his religious conversion and kept him on the "straight and narrow."

Billy describes the way he gave up tobacco: "I went on Monday and got a paper of tobacco. I took one chew out and had no desire. The Spirit of God was striving with me. He was preparing me for His Spirit. He will not put His Spirit in an unclean place." Billy never used tobacco again. He was converted a few days later. He became a clean vessel for the Master's use. "Slobbering 'round" is very repulsive to a clean soul; made whiter than snow, and filled with the Holy Ghost. Billy says, "I do not care to sleep in a cabin where tobacco is. The man who is using it–his breath is offensive to me."

Billy was born in this world in 1834. He was born into the "Kingdom of God" in 1878. His desires and joys and hopes–that had been all sensual, earthly and perishing–became God-like, heavenly, and eternal. Oh, what a wonderful change! Swearing, "slobbering 'round", etc., were earthly and sensual They were dropped forever. One evening, during a revival in the Free Methodist Church, his wife asked him to go to meeting. He asked to be let alone, but she persisted and he went. The sermon deeply impressed him. When he was going to bed that night, he knelt by the bedside and said, "O, Lord, if this religion is what it has been represented unto me, let it be so in me." "I meant it, and it was so. I felt a voice speaking to me. It said, 'Black, arise, thy sins are forgiven thee.' I began to praise the Lord–to shout and to sing. That was the end of my wild career. That ended swearing, chewing and breaking the Sabbath. I used to study all the week how to get the boat on a long level, that I could boat on Sunday. My Salvation ended all–greasing wagons, oiling harness and splicing lines on Sunday. I have been rejoicing in the Lord ever since."

"Immediately I began to tell others what I had received. People said, 'He is crazy'. I wish I had more of that kind of insanity. Some said, 'Wait till the ice goes off the water, and he begins to go boating again. He will lose his religion. The ice has gone off a good many times, and I am a Christian still." When I told Billy that the canalmen constantly say to me, "You cannot be a Christian on the canal," he replied, "If you get Salvation, it will keep you. The Lord is stronger than the devil." Billy said to his wife one day, "Mary, no matter what comes or goes—what troubles come—I will never give it up." That was his spirit. It is the spirit of everyone that is born of God, born into His Kingdom. There is nothing to go back for—nothing but devilishness and misery and hell at least.

I asked Billy how he knew he was saved.

"I knew it," he replied. "It was no hidden mystery. Jesus Christ spoke to me. I knew it. My thoughts were different; my actions were different. What I cared the most for before, I cared the least for afterwards. The taste and desire for tobacco were taken away. The Holy Spirit was within me. If I had not the Holy Spirit within me, I would still be doing the same things. If it were not for the change, you would be the same man still. The Lord does all things well. If the Lord converts a man, you can see it in his face. He has not that cold, dark, ugly expression. There is just as much difference between the face of the sinner and the saint, as between kerosene and electricity."

I said to Billy, "Why were you converted so soon?" (He had only said a few words, and was saved). He replied, "Because my heart was inclined that way. The Lord does not force a man against his will. I longed for it from my infancy. The example that we set before our children is not lost. I longed to get in the way that was taught me. I had a wild nature, but the teachings of my parents restrained me. If they did not keep me from wrong, they troubled my mind. I would get drunk, and at such times I have knelt in my cabin and said, 'O Lord God, this is not the teaching of my parents.' I wanted to be delivered from that which made me a slave. I am fully satisfied, that, if you are willing to be saved, you can be saved. God does not compel. He gives us a good free-will offering. Jesus Christ came, and died, and offered us this great Salvation. If we reject His gift, it is our fault. He did His share, now you do yours. He did His part, now you do yours. If you do, you will be saved. He will not compel, but His spirit will strive with you. If He can save such a sinner as I was, He can save any sinner." At another time he said, "I always wanted to be good—to know God as I had been taught in my early years. I believe my parents' prayers followed me." I believe he received Salvation so easily because he was sincere; he sought with all his heart, and he put God to the test.

The Free Methodists still follow the Wesleyan doctrines. They preach sanctification or the Second Blessing. Billy heard of this doctrine from his wife, who herself had received the Blessing among the Free Methodists. He said, "If there is more, I am going to have it." From Dec. 25, '78 to Sept. 25, '79, he sought for the Second Blessing. He told me, "I sought it nine months and I would not have given up till this day, if I had not received it. I received it. I knew it. I shouted louder than ever. The people said, 'He is crazy'; but I knew what I had. I am glad to be a madman for Christ's sake, and I will be glad to be madder still. The Second Blessing strengthened me, and made me bolder for the Lord. I appeared to be more filled with the Spirit."

I subjoin a few of his remarks which show the spirit of the man.

Speaking about God's goodness he said: "The Lord has always been good to me. I think He knows who will be His. He has His eye upon them–His mark upon them–and everybody can be His if he will."

Speaking about the Spirit he said: "No matter what comes and goes–how they injure you, the Spirit will come round, and you are willing to forgive; willing to do good. You never hold hard feelings. You have no bitterness, no enmity. You forgive as Christ forgave on the cross. It don't matter what is said about you, the love of God in you forgives as Christ forgave."

Speaking of the Christian life, he said: "I have grown in grace, and I have lost nothing, even in my sickness. The spirit of God in me is as fresh as ever; and when the Spirit comes on me, I never quench it. People may laugh. I do not mind. I am so filled with the glory of God that I do not mind them. All I am sorry for, is that the whole world could not hear me." Again he said, "The longer I am in the way, the calmer and milder and sweeter I appear to get."

Speaking about heaven, he said: "Jesus is there. He has gone to prepare a place for me, and I would like to bring others with me and not go empty-handed. The beauty of the Lord's Salvation is, that you are willing to do anything for His cause. If you are not, you have not got it. Even for one sinner I would willingly die. I often have had that feeling." I believe Billy's desire will be granted. On May 25, 1901, he and I prayed with a sinner in his cabin. The man seemed very much affected, even penitent before God. He shook hands with me after our prayers most warmly.

Speaking about the world he said: "I have to make a living in this world. This is all I care for. I do not wish to be a drone, or a sluggard. Riches don't bother me. I am bothered to serve Jesus the remainder of my days. The Christian should not be anxious for a surplus, but he should be for a decent living. If he does not, he brings disgrace on the Cause. They say, 'He is a regular drone." I have noticed in all Billy's talk, that he thinks of the glorious Cause of Christ which he represents. He boats wood on the canal–not coal. I said, "Will you take down a load of coal, if they wish?" He replied, "They have been accommodating to me, why should I not be to them? The more obedient you are to your superiors, the more they will think of you, and it speaks well for the cause you represent. If you are not accommodating they say, 'There is a man who professes to be a Christian, and is contrary'. If you profess Christ, you must be an example of godliness, all these things have to be looked into. You have to be a man of a meek and lowly spirit."

I asked him if he loved the Word of God. He replied, "If I did not love the Word I had not life in me. Christ is life. The Word is meat and drink to the soul. I cannot live without it".

I asked him if he loved Christ after he was converted. His answer was, "As soon as I was converted my love for Christ exceeded all things I ever heard." Again he said, "There are no troubles, no trials that can erase the love of God from your heart. Nothing can, if it is once established in you."

If God's Holy Spirit led Billy Black into the Kingdom of God, He can lead any sinner. If the wondrous grace of God abounded towards the wicked canalman, so that he became a totally new man, God's grace can save any boatman or other sinner. His Word says, "He will have all men to be saved, and to come to a knowledge of the truth." Again the Word says, "Whosoever will, let him take of the water of life freely."

HELENA STONE

Age 95 in 1976. Born at Stewartsville in 1881. Daughter of Henry Stone, operator of a store along the canal. Now living in Mayfield.

Chapter Seventeen

Helena Stone

"We used to fish out the store door, the feed door, by the side of the canal. I loved to fish. I don't think my sister ever did. We used to catch catfish and perch and we loved eels."

The following is a conversation taped between Mr. James Lee and Miss Helena Stone—December 15, 1974.

Now what is your full name?
Helena W. Stone.

And how old are you, Miss Stone?
I was 94 last September 16th.

What has been your connection with the Morris Canal?
Well, my father owned a store beside the Canal in Stewartsville, and also the coal yard. We had pea, chestnut and stove coal.

And you got this coal from the canal boats?
Yeah, the coal was brought there by canalboat from Phillipsburg. I think that's where they loaded.

How did they get the coal out of the canalboat?
Well, my father had a long tall pole; I think you'd call it a derrick. There was a rope with a pulley on it. A large bucket was attached to a rope on a horse that pulled that rope; went foreward and pulled that bucket from the barge. There it was emptied on our coal piles and of course all that coal was brought during the spring and summer for the winter supply for those who bought coal from us.

How many boats would you say your father got?
Well, I imagine at least seven or eight.

Well, there was 70 tons to a boat. That would be a considerable amount of coal. How long did it take to empty a boat, do you remember?
Almost all day. The boys and men would help him lived, I suppose around Stewartsville.

What was your father's name?
Henry H. Stone.

And your mother's name?
Lilly Wilson Stone.

Were they life-long residents of Stewartsville?
My mother was born at Still Valley.

And her maiden name was Wilson?
Oh yes.

And your father—was he originally—
He came from down, can't think of the place but it was along Bloomsbury. He was brought up on a farm. If I could find those pictures I took. I was called an expert, I guess, on taking pictures. My father made a dark room for me up in the store, upstairs and I finished my own pictures.

Did you have any relation on the canal?
No. I don't think any of them.

Your father had a store along the canal. Did many of the boatmen stop there for their supplies?
There was a stopping place. If they were near there at night they laid in their supplies, on their way to Paterson.

Henry and Lilly Stone sit on the porch of their home in Stewartsville. Cigars and tobacco were a popular item for boatmen.

What might they buy there?

They laid in their supplies at my father's store. For their trip back home—on back to the starting point.

Did your father have a barn there for the mules?

Yes, a mule stable and if they needed feed for the mules. They'd bought feed from my father. He had what was called a feed door in the store on the canal side that opened. And that's where they loaded and unloaded their feed. Those bags of feed for the mules.

Do you know where they kept the feed on the boats?

No.

Well, it was right on the hinge in the center. There were two big boxes there. They could pull the boats right up along your father's store and hand it out the door. That was very convenient for them.

Did your father extend credit to the boatmen?

Oh yes. We had what we called Store books and everything we put in that, I guess if they didn't have the money till they came back. He trusted everybody.

How did that work then. Did you keep a book and they kept a book?

Yes.

And it was wrote down at the same time in each book what they bought.

They had store books; I guess they called them ledgers. I never had anything to do with the books. I had really very little to do with the store. My father used to say that I didn't know how to tie up a pound of sugar. I said I don't want to. I never clerked in the store. I never helped in the store, unless it was at Christmas time.

That was a general store?

That was a general store.

What all did you carry in that store?

Everything from a needle to a plow.

And also dry goods?

Dry goods and groceries. One was paint. We sold to most of the painters around there.

I'm interested in the children too on the canal boats. Do you remember—were there any small toys you carried for children in your store?

We sold toys for Christmas but I don't remember the boatmen.

Dolls and whistles?

Father sold dolls, very lovely ones. He went to New York twice a year with my Uncle Oscar Wilson. He stayed overnight. He bought goods. He'd go once in the spring and about two months before Christmas he laid in his Christmas supplies.

How were your supplies delivered to you from New York?

By railroad. My father would go to the freight office.

Did your father sell furniture?

He took the young people into New York when they were married or afterwards or before and let them buy their furniture for housekeeping. Select what they wanted and then it was sent to Stewartsville by freight. They picked it up and paid father as they could. And I don't think he ever lost anything that I ever heard.

Was your store ever robbed?

That time that my father found in the morning there was robbers but they didn't get anything.

They didn't take anything at all?

It didn't amount to much.

They broke in?

They broke in. That's the time he walked down the canal and tried to catch them. He tracked them in the snow.

Did you have any dogs to guard your store?

We had two dogs—a bull dog and I don't know what the other one was. But I guess the dogs—they either didn't care anything about the dogs or else the dogs didn't bother them.

How did they break in?

They cut the bottom panel of a double store door. We had two big bay windows in the store; one on the left and one on the right of the entrance. Those doors or shutters were put on barred at night. Every night they were put on and had a heavy iron bar that bolted them.

And that was the only time it was robbed?
The only time.

Did he ever find out who did it?
No.

Did he go after them?
He went after them in the morning. He saw the marks in the snow. It had snowed a little and Mother said afterwards (he didn't take his gun or pistol with him)–She said, "Why didn't you ask the robbers to wait till you got your gun or your pistol?"

And how far did he follow them along the canal?
Almost down to the next plane–#9 plane.

Port Warren and there he lost the tracks in the snow?
There were too many. I suppose they mingled with somebody else's.

Did your father hold an elective post in the township?
Yes, he was tax collector until he died. The last year he wasn't able to go to the town meetings to report. I can't think of the

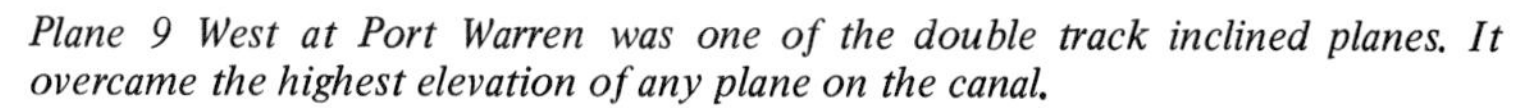

Plane 9 West at Port Warren was one of the double track inclined planes. It overcame the highest elevation of any plane on the canal.

lawyer's name but he said "That's all-right."

You could do that then?

Yeah.

How many hours did your store operate?

Well, Father went over about half past seven because many of the men that worked on the railroad would stop and get tobacco and whatever they wanted on their way to work.

And when did it close at night as far as you can remember?

I think he closed at 9 o'clock at night.

You said you father went over. Apparently you didn't live at the store?

No.

Where did you live in relation to the store?

We lived . . . the scales where they weighed the coal was between the house and the store, and it was as far as from here to across the street. We weren't connected with the store and we were the first ones to have steam heat. My mother didn't want heat in the house for fear it would interfere with our keeping things in the cellar. So Father built a furnace room, dug out half-way between the store and the house property and put the furnace down there and run a pipe into the house and we had a wonderful furnace.

How about inside plumbing?

He put it all in himself. Father put a bathroom in. We had the first one in Stewartsville.

The first inside bathroom?

Right.

There were plenty of outside ones though.

Oh, yes and we had an outside one too. Father was very mechanically inclined. Of course we didn't live to have an automobile. They weren't in existence.

When did your father die?

He died in the spring of 1911.

Then he was a pretty young man when he died.

Yeah, I think he was in his 60's.

The store of Henry Stone at Stewartsville carried general merchandise and served everyone. The windmill was for the water supply.

Did the store operate on Sundays?

No. Father always made us remember that people had to work seven days a week and he never wanted us to buy anything on Sunday. But if they were out of bread or any necessary food he would go over and get it out for them.

He was real accommodating then. That's why he enjoyed a good deal of business, I suppose.

That's right.

Did you have any brothers or sisters?

I had one sister who died when she was fifteen.

No brothers though?

No. My mother and sister died the same week. My sister had pneumonia.

What do you remember most about the canal?

Not too much because I had very little to do with the canal.

Did you ever swim in the canal?

No. My father and mother was afraid we'd drown.

Did you ever fall in the canal?
No.

Did you ever throw stones in the canal?
Oh, sure.

Do you remember anyone falling in the canal? Drowning or anything?
Our horse fell in the canal, but they got him out. I forget how. I think they put a belt under him and lifted him out.

Horses don't mind falling in as much as mules do. Mules panic, but a horse generally doesn't panic. Do you remember anyone in Stewartsville ever drowning in the canal when you lived there?
Yes, there was a young boy by the name of Steck. He was visiting his grandmother, I think, and grandfather. I forget their names too. He drowned.

Did you ever go ice skating on the canal?
I never did, but my sister used to skate from Stewartsville to Phillipsburg. She was a wonderful skater but my ankles were too weak.

Did anyone ever play any jokes on your father?
Yes. My father was a great joker. Too much so. One night I had a friend visiting me for about a week, I guess. I belonged to what they called the Goodship Fellowship Club. Couples from Phillipsburg and around used to come out and that night after the party was over, we'd gone to bed and we heard all this—what did they call it in those days—I can't think . . . calathumping! And they came and had tin cans and dish pans and I don't know what all, and we woke up to this noise outside of father's window. This friend and I couldn't imagine what in the world was going on. And it seems a couple of days before, father had bought two or three ducks for us to cook at this party and some man was in the store heard him ordering and he said, "What's all this?", and the day before this man in the store said, "What's all this brine for?", and father said, "I'm going to get married, don't you know?" Of course it was no such thing and the next day our

minister came over. Oh, father said, "This is crazy." He said to the group outside, "It's only a joke." He said, "I never intended to get married." But if anybody asked him anything he was just as liable to not tell him. I think I inherited it from my father because if anything was told to me, I usually kept it to myself rather than spread it. The next day our minister came over. He had heard it and father said, "Are you crazy?" He said, "Why that was only a joke."

The minister too thought he was getting married?

I guess. And I said to father, "Now father, this is one time when you've gone a little too far. Why did you tell him that?" He said, "They had no business to be so curious." That was one of the jokes.

Calathumping. I hadn't heard that in a long time. When somebody got married, they'd come around and serenade them and they'd give them money: nickles and dimes.

Oh, they wanted father to come out and treat them to cigars. He said,"You're crazy." That was only a joke." I think it was some men up in Low's Hollow.

Yes, pretty tough neighborhood up there.

Father just turned and said, "Don't you know I'm getting married!"

Maybe it was one of the Unangsts up there. They're a pretty tough lot.

Might've been—yes they were.

There's a lot of them too.

I should say so.

Do you remember any songs you might have sung as a young girl?

No, I had no voice. I was no singer. I don't remember any. Of course we sung the Christmas Carols. Oh, my father used to play violin by ear. He was never taught but just naturally. My mother played the melodeon in a Lutheran Church for the choir.

Did you ever hear the boatmen play any music as they passed?

Oh yes! They had accordians.

Yes, and violins and banjos.

Oh, at night we could hear the banjos.

Did you ever hear any noise from the canalboats when they were approaching a lock or inclined plane?

No, it was a little too far from us.

Did you ever hear them blow that conch shell?

Yes, I remember that. Of course I went back and forth to Centenary in Hackettstown, CCI, it was then, and I wasn't home too much except at night. I traveled by train. Then I went to Easton every Saturday for piano lessons before I went away to the New England Conservatory of Music.

You were kept pretty busy then and weren't around the store in your later years. Now Centenary is a girls' school but at that time it was a co-ed school.

That's right.

Do you remember what year that was?

Well maybe we can trace it back. I was about 11 years old.

So that would have been around 1891 or 92. How many years

Canal regulations required a captain to signal a lock or inclined plane of his approach. The conch horn could be heard over a quarter of a mile.

did you go there?

I went there about a couple of years and then I went to the New England Conservatory.

Was there a reason why you went there rather than a local high school like Phillipsburg?

Yes, my father was on the school board.

What school board was that?

Stewartsville, in Warren County and our teachers he didn't consider were, I guess too good.

He felt you would get a better education going there.

That's right.

Do you remember any famous neighbors?

The Thompson's—Will Thompson was station agent and he was a great friend of my father's. In fact before my father moved to Stewartsville, before he was married he boarded with the Thompson's family, right next to the station. We were very very close.

Do you remember a family by the name of Brill?

Yes, they had a cattle farm. I think just below our bridge, as people called it, not quite down in the town.

They raised prize cattle down there. People came from miles around just to buy their meat.

Right.

Do you remember between your store and plane 9 where I live there was another bridge. There was a bridge the farmers used to get across the canal and I understand the boys used to go swimming there.

That's right. What did they call that? That wasn't too far from us. We never went down.

The boys used to go skinny-dipping there. Is that true?

They did. That wasn't a mile from us. I think they called it "Someone's Swimming Hole," but I don't remember.

But you do remember it being there?

Yes.

Is there anything else that stands out in your memory?

My father was a Mason–not by trade. He belonged to the Masonic Lodge.

What town?

Phillipsburg. Also the Odd Fellows.

How about the Junior Order of United Auto Mechanics?

No. He was made an honorary member.

How about the Grange?

No.

When he went to Lodge meetings did he dress up in a fancy uniform sometimes?

Yes, yes–the Tall Cedar's too.

And your mother was an Eastern Star member?

No, I am. In fact I belonged for more than 50 years.

Do you remember any black people living in Stewartsville?

There was one family down at the end of the town, but I never knew them very well and they came to our store.

When you say the end of the town–

Towards Bloomsbury. You know, there was a road that went straight through, I think to Bloomsbury.

Low's Hollow, up that way?

No.

What was their names, what did you call them?

I never knew their names. I just remember that they traded at our store. I don't know where they went to school. I just remember there was a colored family down there somewhere.

Now were there any other colored families?

There was a colored family that lived in Low's Hollow, up beyond that Methodist Church. Their name, as we knew them as children, was Aunt Ann and Uncle Brewster.

And did they trade at your store too?

They traded at our store. They used to walk down and we said that Uncle Brewster would have a birthday every month because he would come down and say to my father, "You know I have a birthday this week." Father would say, "Oh, you do? Well then, I guess I'll have to give you some tobacco."

He got tobacco every month then?
I think so.

Did he work anywhere that you know?
I don't remember. Mother used to fill a basket full of a picnic lunch and take it up and we'd have lunch with them. Of course, they didn't call them lunches. They called them dinners. And we'd have our dinner with them and then come back.

Do you remember that little church up there in the Hollow?
Yes, I do. My father used to give every time they'd had a harvest home dinner (which I think was every August). Father used to give them a barrel of sugar and different things for their dinner.

Did they have a regular minister there or did he come around, circuit minister?
I don't think they had a regular minister.

It was a very small church.
I don't know where he came from.

Sometimes they would only have church every other Sunday in some of these small ones.
That's right.

And sometimes they'd have one in the morning and another someplace else in the afternoon.
They always had benches and seats if you wanted to go up there and have a picnic, you could. There was a spring above there and there used to be a man who we always thought was sort of odd who lived near that spring. We weren't afraid of him but some people were. His name was Swietzer. He lived there all alone. He was a hermit.

Do you remember any other industry in Stewartsville? Now at that time there must have been a carriage shop.
No.

Do you remember a family by the name of Wellers? They were quite a few Wellers. They had their own cemetery, between Stewartsville and New Village.
Oh, yes. There was a Mrs. Wellers that was related to Mrs.

William Thompson.

That would be the wife of the Station Agent?

That's right.

Did you ever go fishing in the canal?

We used to fish out the store door, the feed door, by the side of the canal. I loved to fish. I don't think my sister ever did. We used to catch catfish and perch and we loved eels. My cousin from Phillipsburg, Waterstone, used to love to come out to fish. He was I suppose, about 12 or 14, and he never could catch any. He was very fond of my mother. He used to come out and spend the day from Phillipsburg with us and one day he came running in and said, "Oh, Aunt Lilly, I caught a fish, I caught a fish!" and she said, "Oh you did?" and smiled. My father had tied a salt macherel to the end of his fishing line and he pulled it in.

And he thought he caught it?

He must have been quite young, you know, or he would have known.

That was probably the only salt mackerel ever caught in the Morris Canal.

I guess so!

Were there any famous men that came to Stewartsville?

Well, Thomas A. Edison. My father sold coal—the first coal ever used at the Edison Cement Plant. We had two horses and a big coal wagon. I have a picture somewhere but I don't know where it is. And Thomas A. bought the coal from my father and father and the boy (we had a young man working for us) used to take the coal up. I don't think father ever went on that order. The boy used to take the coal. Father sold his coal to Thomas A. Edison. In fact he sold him, I think, all the coal they ever used there. I don't remember about the later years because the Edison closed, I don't remember how many years the Edison Company was there. But my father used to go out when Thomas A. came down to pick up eggs that he'd bought to bring down to Orange over the weekend. He and my father would have quite a long talk out beside the buggy

The side door of Stone's store through which boatmen loaded feed and children fished is open. The boom is for loads of coal.

that Thomas A. drove from New Village. And he would take home eggs and I suppose butter because we took in butter in exchange for food, you see.

Oh, people would make butter and bring it to your place and they would swap food for it?

That's right. I know eggs were about 10 cents a dozen. I don't know how much butter was—very cheap and then they would pack, my father would pack stone crocks full of butter and the boatmen would take it on their boats for their trip back home.

Is that where the term tub-butter came from?

Yes—I don't know. You see, they brought it in rolls and sometimes they would have all kinds of designs on these rolls of butter.

Then your father would weigh it and determine how many pounds?

If it was good. He would always taste it, and if it was good, he would pack it for them.

Did Thomas Edison come there quite a bit then to the store?

I think almost every week before he came home. I never met him. I never went over. Father talked to him. They were quite close.

What did the inside of your father's store look like? Can you picture in your mind what it looked like as you walked in the door?

On the right hand side, of course, was a bay window. As you went in, they had bins, one bin was for coffee. They were pretty good sized bins behind the counter. Oh, we had a long counter. That was on the right and father used to weigh the five pounds of sugar there and have it ready when people wanted it. Once in a while at Christmas time maybe I'd go over and help a little. Father would say, "Oh, you're no good tying up groceries. Nobody would ever know you were a storekeeper's daughter." I said, "Well I never want to work in a store."

Were the hardware and paints in the back of the store?

That was upstairs, 2nd floor. The feed room was in right in the back, facing the canal.

Where was the candy case located?

Well, they had candy jars and we were never allowed to go in and help ourselves. In fact we didn't have much. One thing at Christmastime the candy was crystal clear and there were all kinds of designs it was made into, like a little engine or animals, and different things.

On a stick?

No, all separate. It was clear candy. He had jars that he would keep peppermint sticks and two or three other flavors. But we sold some candy—not much. But at Christmastime my father sold the most beautiful dolls. He went to New York on those trips and he bought beautiful dolls. They were in a box separate. People came and bought them.

Now, Miss Stone, you know we are taping this. Do you mind using this for public dissemination in any way?

I don't know why.

It's for the State of New Jersey and we want to record your memories as far as you know of your father's store and it is something that will be left for future people to read and study.

It's all-right with me. You know I should have told you, I guess, that father sold everything from a needle to a plow. He sold hardware and paint and dishes.

Horse collars?

No, he never sold them.

Never sold harnesses?

No, he never sold harnesses either.

For that you'd go to a harness maker.

He did invent and sold—I can't explain it—it was something that would hold a trace to the side of the shaft, isn't it?

Yes, the shaft went to the traces.

Well, those traces were long and they used to slip off sometimes and he invented something and made and sold something that held the trace—the leather strip to the shaft. He never made much of a success of it but he did sell some.

Well that's interesting too. You mentioned your father had running water in the house. Did he have a pump of some kind or what?

He had a windmill up in the attic of the store.

Did he build this windmill himself?

That I don't remember.

Where did he get it from?

I don't know. He piped water for the store and for the house.

Well, did he have a well drilled then or did he get the water from the canal?

The water came from the canal.

But you didn't use that for drinking?

No, not for drinking—we had a cistern.

You trapped your rain water from the roof?

Right.

The Port Delaware loading chutes were just one place boatmen would socialize, often holding informal musical sessions.

James Lee listens to Edward Lenstrohm as he plays "Go Along Mule," at Waterloo Village during the filming of **"Famous Tiller Sharks"** *for the New Jersey Public Television Corporation.*

Chapter Eighteen

Canal Songs

There were many songs sung by the boatmen and their families on The Morris Canal to relieve the boredom and drudgery of everyday living. Sometimes they would sing as the boats were moving along, but more often on a Sunday when the canal was closed and the boats tied up for the day.

Occasionally they would take a popular song of the day, or a hymn and add their own words to the tune.

Unfortunately most have been lost with the passage of time, but here are a few that have been saved:

"Go-Along-Mule" was a popular song on the canal, and it is sung here by Edward Lenstrohm, who was taught the song by his father, Captain Peter Lenstrohm.

GO ALONG MULE

I've got a mule, she's such a fool,
She never pays no head.
I'll build a fire beneath her tail,
And then she'll show some speed.

I drove down to the graveyard once,
Some people to visit there,
But when a black cat crossed my path,
It sure did raise my hair.

> Go along mule.
> Don't you roll those eyes.
> You can change a fool, but a doggone mule
> Is a mule until she dies.

Henrietta found a hen,
Don't know whose hen she found.
She's such a good hen, she lays an egg,
But Henry, she lays around.

A Man in Boonton pulled a gun,
And took a shot at me.
But when he took the second shot,
I passed through Port Moreee.

Go along mule.
Don't you roll those eyes,
You can change a fool, but a doggone mule
Is a mule until she dies.

I bought some biscuits for my dog,
And put them on the shelf,
Times got so hard, I shot the dog,
And ate them up myself.

I'm going down to the river now,
To lay me down and die.
But if I find that the river's wet,
I'll wait until it's dry.

(CHORUS)

Go along mule.
Don't you roll those eyes.
You can change a fool, but a doggone mule
Is a mule until she dies.

FAMOUS TILLER SHARKS

"Famous Tiller Sharks" is a poem written by Charles Matlock Hummer in 1959 many years after he left the Morris Canal, but in his fading memory, many of the faces of his former buddies and acquaintances of the canal would reappear, so he composed a poem to honor them.

This poem was first used by James Lee in his book "*The Morris Canal: A Photographic History.*"

It was also used by the New Jersey Public Television Corporation in their film *"Famous Tiller Sharks"* and put to music by Byron Bernheim for the television presentation.

Remember Frog Mouth Patsy
And Paddy the rat, as well.
Jimmie the mouse, old Paddy's son,
And Paddy's good wife Nell.

Remember the poor old Dagon Brothers,
The Pendys and the Peers,
Pigeon Nixon and his gang
Of Jackey's racketeers.

Must stop my rambling
This sheet just not afford,
The Famous Tiller Sharks
That rode the water board.

The Turn Boys from Easton,
The Blacks, both Bill and Lon,
Pug-nose Jim with the boxed-up chin
And Guinea Hollow John.

Charles Matlock Hummer was born in Oxford, NJ, on November 4, 1887. He was the son of Captain William Hummer. He died on November 7, 1959.

Then Big Bill Gardin,
And Harry Cisca Bublin
Their middle was so very wide,
They could hardly squeeze in the cabin.

  Must stop my rambling,
  This sheet just not afford,
  The Famous Tiller Sharks
  That rode the water board.

McLouton's Irish Steam Boat
Stamped "Hit or Miss We Go".
The Healys and the Tuckers,
The Ruggs from the overflow.

Gallant old Theo Lewis,
And his faithful better half.
Bill and Amos Stackhouse,
The Jewell boys all araft.

  Must stop my rambling,
  This sheet just not afford,

The captain had to be constantly on the alert. The tiller had to be moved as the canal twisted and turned.

The Famous Tiller Sharks
That rode the water board.

Happy old Jim Camel,
Jim Haines as happy still.
Long-fingered Lew, the bug-a-boo,
And giggling dark-skinned Bill.

So hold the channel captain,
Till we meet those gone before.
At the everlasting snubbing post,
Along some happy shore.

(CHORUS)

Must stop my rambling,
This sheet just not afford,
The Famous Tiller Sharks
That rode the water board.

Some of the famous tiller sharks: (left to right) Beecher Dagon, Teddy Dailey, John Dagon, and Sam Peer.

GOING DOWN TO COOPER'S

"Going down to Cooper's" was a true folk song of the Morris Canal. Cooper's Furnace in Phillipsburg was one of Peter Cooper's early ventures in the iron industry. The people in the song were all from the Phillipsburg area.

It was a fun song that was fondly remembered by the people that were on the western end of the canal.

Going down to Cooper's just six o'clock,
Who did I see, but old Aaron on the dock.
And he said me jolly driver; Now whose team is that?
Sure 'tis old Mike Cavanaugh's, just a getting fat.

> Fal-ra-di-aiedo, Far-al-de-ya.
> Fal-ra-di-aiedo, A-hum-de-dally-dae.

The driver was lame, the shaft mule was blind,
The lead mule had a corncob sticking out behind.

> Fal-ra-di-aiedo, Far-al-de-ya.
> Fal-ra-di-aiedo, A-hum-de-dally-dae.

Pete was at the hinges, Patsy at the bow,
Mike was at the tiller handle, showing them how.

> Fal-ra-di-aiedo, Far-al-de-ya.
> Fal-ra-di-aiedo, A-hum-de-dally-dae.

Going down the "17er", we were doing mighty well,
When the mules broke the towline, the boat went to Hell.

> Fal-ra-di-aiedo, Far-al-de-ya.
> Fal-ra-di-aiedo, A-hum-de-dally-dae.

"Hunks-a-go Pudding and Pieces of Pie" was a ditty that the children used to sing, as they walked along the towpath behind

the horses or mules that were pulling the boats.

Mrs. Chester Mann, the daughter of Captain Peter Lenstrohm, remembers it well, for she used to sing it walking behind her father's mules.

HUNKS-A-GO PUDDING AND PIECES OF PIE

Hunks-a-go pudding and pieces of pie,
Me mother gave me when I was a boy.
And if you don't believe it, just stop in and see,
The hunks-a-go pudding me mother gave me.

"You take the hatchet and I'll take the saw" was another ditty that Mrs. Mann remembers. Although it sounds kind of gruesome, it brought smiles to the faces of the children who sang it.

You take the hatchet and I'll take the saw,
And we'll saw off the legs of me mother-in-law.

Mr. Russell Batson remembers a ditty that one boatman would sing to another as he would overtake and round a slower boat on the canal.

It was all in good fun, and the slower boatman accepted it in the good spirit for which it was intended.

Oh, you rusty old canaller,
You think you're mighty nice,
Standing by the tiller blade,
Picking off the lice. PSSSSSSST.

These are only a few songs and ditties that were once popular on the Morris Canal.

Hopefully more will be found, as interest in the old canals seems to be increasing.

ARTHUR UNANGST, JR.

Born at Low's Hollow Road near Stewartsville. Died at Sussex in June 1976. Son of Arthur Unangst, a master carpenter who worked forty-six years on the canal.

Chapter Nineteen

Arthur Unangst, Jr.

" 'I am the Devil!,' Sam said. The old man said, 'By Jesus, if the Devil don't look any worse than you do, I ain't a God Damn bit afraid of him.' He went into him, he said, and knocked the Hell out of him!"

An interview with Mr. Art Unangst by Mr. James Lee

How old are you now, Mr. Unangst?
79.

I have got this on tape. You don't have any objections to my taping this, do you?
No, I don't care. I don't care 'cause I never do anything I shouldn't. I used to, though.

Mrs. Unangst: Don't you go swearing on that tape now, Dad.

Oh, we can stand a few words.
Mrs. Unangst: You know what? If you do, it'll come right up from there.

I don't swear anyhow.

You were telling me about your father. Did you ever boat on the canal?
No, I never did.

You never did boat, but you helped your father?
I rode with him. He used to take me with him just for a ride.

He was a carpenter boss?
He was a carpenter himself.

Did he have men working under him?
Yes, he had men helping him. He would build the bridges on #8 plane. He would frame them and then he would put them on the scow and truck them to where he wanted them. It didn't matter where it was on the canal. Then he would put them up and frame them. I would go along with him sometimes.

Green's Bridge House, near Lock 8 West, was one of the many hotels along the canal that catered to the thirsty boatmen.

But this time when he was down to Green's Bridge, I was going to tell you, he was in Green's Bridge and he was in the hotel and this Sam Bowers was in there and the hotel keeper and Sam Bowers got to talking about this old man. Sam Bowers was a husky man, you know, and a nasty guy. The hotel keeper said, "Art Unangst, why, he would tear you apart." And this made Sam mad. The old man had had a few drinks, I suppose, and went over and was laying on this little bench there in the back part. So the hotel keeper and old Sam went and old Sam grabbed the old man from the back of the neck and the seat of the pants and threw him right out of the shed there. They said the old man up and wiped his eyes like this and looked up and never had seen him before. He said, "Who are you?". "I am the Devil!", Sam said. The old man said, "By Jesus, if the Devil don't look any worse than you do, I ain't a God Damn bit afraid of him." He went into him, he said, and knocked the Hell out of him!

Apparently there was a lot of fighting on the canal.

Yes, there was.

Did your father ever tell you about any of them?

Oh yes. He told me about it. He said one time they had the

scow. They had a break on the canal and he went down in and was helping them. Uncle Hen was the boss there, you know; that is his brother. So the old man went down there and was fixing the break and bringing mud up from down over the banks. Along comes a guy who wanted to talk to Hen somewhere, and they got into an argument. The old man come up and looked at them and went back down in again. Pretty soon he come up and there where that mud had flowed down from the fields over there near New Village. The old man up and decided he would hit him and he knocked him right over–ass over–right down. He slid down in the mud, slid right down to the wheel in the mud. And he come back and they had an argument again and up come the bigger boss, Aaron Vough, and wanted to know what was the matter. The old man went right back down in the hole again. He didn't want anyone to know anything about it. Just because the man was arguing with Hen, he put the works to him. He wouldn't take nothing from nobody. Never did.

And he wasn't too big either in size?
Oh, he was a big man.

But he wasn't real big.
About my size, 185 pounds and 5'6"–somewhere around there.

He had other brothers working on the canal, didn't he?
Oh yes. He had Barney and he had Uncle Bill. Uncle Hen and Uncle Parmeton and Uncle Os worked on there too. Fred, he lived in Stewartsville. I believe he is dead now. He was a young guy at that time. He had John Unangst working there, I know that.

Were any of them boatmen or did they all work as carpenters?
All worked around boats and the scow and like that. Frank Piatt was the underboss. He was under Aaron Vough.

How about Isaac Durling?
Durling, I heard him talk about Durling. He was a higher boss, I believe.

He was like the boatmaster. He was at Phillipsburg and he made

sure that the boats were well-equipped when they went out; that they had an extra length of rope and that they had the supplies they should have to make the trip.

I think Mutchler worked on the mud-digger.

He might have done it occasionally, but he was also a carpenter and he lived in Broadway. He had a gang and the carpenter boat would go from one place to another. I talked to a fellow by the name of Mr. Charlie Snyder from Stewartsville. He only worked from when he was age 18 until he was 21 on the canal, and then he left it. But Mutchler was his boss. But, oh, the stories he told me about your father! You know, nothing rough, you understand, but he didn't take anything from anybody.

Yeah, but he done a lot of that to be devilish you know too. He done that to have a lot of fun. Funtime! Dory Metz, he had that tavern down there. One time when I was only a young kid, and I didn't take nothing from nobody either.

Like father, like son.

And I was down to the Wagonhouse and down comes Brick Searles. I was just unhooking my horse coming from Brainerd's. See, I worked in Martin's Creek then driving horse back and forth. The horse was wet and I had to unhook the traces and he says, "Don't unhook yet. I want to go to Stewartsville and get a bottle and a box of beer." I looked at him and I says, "All-right." Bill Weller was with me and we all three went on to Stewartsville. We got the box of beer and the liquor and tobacco and I stayed in the Wagonhouse there. We got to drinking it up, we did, between us three. Mom kept coming down all the while and hollering to come for supper, but I didn't want no supper. I put the horse away and took care of that and we drank it up and along come the Midnight car. He says, "Here comes Slim and Beef and they are each going to get a quart and I am going over to see them." So he went on over, went across the bridge and Archie Meyers was coming there, Tyne Metz and them. "Hi. Hi. Hi.", they went. They got over there and met Dory. By God, I heard them hollering and they got in a fight. I went on up, like a fool, I went on up and I says, "What's the matter with you guys? What are you scrapping for?" Dory come around and hit me and knocked me right across the road. I went back and went at him. He didn't ever hit me once again. I put the works to him and he run. I blackened his eyes as black as coal.

He was a big man, too.

He was a big man and he went across the bridge like a horse and his brother said, "What the Hell town you call this?". That was Tyne. There was nothing to it. The other guy was sitting on the bridge. They seen Dory and Bill, they all seen that. Mom came out, by gosh, came out in her bare feet right in the snow. She heard my voice and she came running down there and says, "Come on, go in the house." We weren't fighting then, but I was after my flashlight and I said, "I'll be in pretty soon." I am sorry to think that she had to get out of bed and come out there like that. I had to go to work in the next morning–that was Sunday. So by gosh, I went to work and Pop had heard something about us being in a fight. He had asked Mom how it started and she said, "He didn't

The Morris County Traction Company's trolley ran under the canal near Ledgewood, part of the line from Dover to Lake Hopatcong.

start it but he wasn't going to stand around and have someone knock his teeth out and take it." So he went back out and said, "What?". He thought he had knocked my teeth out, you know, so he went up and called Metz and everything. He said he'd whip the whole bunch of them. So that night when I come home from work, I walked in and he was eating supper and he sat back in his chair and looked at me. I said, "What are you looking at?". He said, "Open your mouth". I said, "There ain't nothing wrong with my mouth". Pretty soon I opened my mouth and he said to Mom, "I thought you said they knocked out his teeth." And that was that.

Now Bobe, my brother, was living at that time when Dory started the fight. I went down to my father's, he was using that little place there by Shillinger's Mill. He was talking to the old man. I went in and the old man says, "Go up and get a little beer." So I says "Okay." So I went up. Bobe had been up there and on the way back, Bobe said to Dory, "Dory, this is my kid brother!". Dory had always said, "I am going to get him if it takes the rest of my days," but he didn't. He couldn't even see. His eyes were black. He says, "I'll get him," but he never did get me.

Those Searles, there were 3 of them–Brick, Slim and Beef. Were there any others?

They are all dead now, but one. Brick is still living yet.

They used to live in that old shack with their mother?

Yes Sir–up there with Carrie. You know he never drank a drop while they were living. He used to beat them up because they drank. After Carrie died, then he got on to it. Now he is nothing but a drunk.

Somebody had to take care of them, I suppose. Now he takes care of himself. Didn't your father ever want you to go on the canal?

No. I started to work when I was a kid at the Ingersoll. I was only 14 years old when I started to work there. Mom needed the money, she said.

Did your father ever tell you any experiences where he almost lost his life?

Yes Sir. He did! He told me down there to the boatyard one time he was working on a bridge there. He lost a pinch bar. He dove in there and he got in a whirlpool and by God, he said he was pretty near gone. He could swim like the devil, too. He said he made a dive and he got out. He had the pinch bar in his hand, too!

You would wonder what would cause the whirlpool in the basin.

You would, but it was a whirlpool. He couldn't get out. That is the only time he told me he pretty near lost his life.

The canal didn't work on Sunday. I understand it didn't work on some holidays. Did your father ever work on some repairing though when the boatmen weren't working?

No, he was off too. Once he was working down to #9 plane. He always wore a white shirt—most generally wore a white shirt. He didn't drive the horse this day, but come a-walking up from there. They lived right below #8 Plane. I think there's 2 or 3 houses there.

Yes, on the right side there going up the plane.

So, Mom said she would wait to see if he came up and then she

John and Elizabeth Unangst were one of several Unangst families living at the foot of Plane 8 West, Stewartsville.

would put the meal on the table. So she didn't see him coming and she said he didn't come and didn't come and didn't come. Well, she said she went out and went to the gate and looked up the towpath and looked down the towpath. She didn't see and she heard a noise so she went through the gate and went out and looked down the canal. There him and a nigger was in the canal. This nigger buck was coming down the canal and the old man buckled. They went right over the bank right down in there and the old man told me, "That nigger could swim like Hell. I got ahold of him, pulled him down in the water and got ahold of his ear and started chewing! He started to holler that I was trying to drown him!" And Mom, she would holler. She saw the whole thing, ya know. On the other side was Mellick's Peach Orchard. She said this nigger got away and swam and swam and got away. He went across the peach orchard and he run this way and run that way. He got away from the old man.

I was wondering too, you didn't want to go to work on the canal?

No. No. I had my job at Ingersoll and Mom seemed to need me to come back. You know I started work there though and I built up fast. She needed me to come back so she would get board. I gave her my paycheck right on until I got married.

Do you remember your father singing any songs on the canal?

Any what?

Songs . . . Songs.

Songs! *You Rusty Canaler, You'll Never Get Rich*

Were there any others that you might remember?

No. He told about a boatman a-coming along. He called at the boatman by name that was on there. And he said that he seen these chickens over there by this farmhouse, you know. He pulled the boat over alongside and this guy he jumped off and he runned the chickens. The farmer come out and hollered, "Hey! God Damn! That is one time I caught you." "Caught me, Hell! The chicken flew off the boat and I am trying to catch it." The farmer jumped right in and helped him catch the chicken.

Helped him catch his own chicken.

Yes, helped him catch his own chicken. And there was a man by the name of Camel.

Jim Camel. He was a colored man.

I don't know what he was, but Camel, he said he was nasty. Cole was going to pull by and Camel said, "You pull by me, I'll cut your towline." Cole says, "If you do, I'll put the pike pole through you." About that time, he must have got off the boat somewhere or other and he, Camel, took this pike pole, see, and about that time the old man come up the towpath. He always did walk fast, you know, and Camel took him for Cole and he wheeled around and the old man hit him. Camel said, "I have been kicked by my own mules and had rather been, than hit like the one he gave me." Cole told me that himself in Washington.

The old man would always borrow a mule from the canal and take it up home there and use it through the winter. "Old Jim" was 32 years old. We used to haul wood. He had the wood business there, and he used to sell wood and he would sell it for $1 a load. Now that would sell for $60 a load.

It is pretty expensive today.

He would take me up there and I don't dare get off that mule's back. I rode him down the canal, because if I got off his back, he would chase me. He would chase you like a dog. I was only a little kid and my behind would get so sore, I could hardly sit on it, so I didn't get off. When I had him there at home I could get by him if I'd sneak by him and when I did get by him, I would get ahold of him and there I wasn't afraid of him, but otherwise . . . The old man said one time "Go out there and fetch 'Old Jim'." "I won't go out there", I said. "He won't bother you," he said. I said, "What do you mean, he won't bother me?" There was a wooden gate there and he took a peg out of it there where you hang the harness on there. He was walking out there and Old Jim did turn and started after him. The old man hit him and knocked him right down in the brook.

How long did your father work on the canal?

46 years.

When did he start?

I don't know.

Do you know when he retired?

No, I don't know. He worked until the canal was abandoned.

Do you remember what his salary was?

I don't know that. I know after that, you see his hands got all crippled up, he could hardly hold a tool, but he worked. He built Oss's barn up on the hill, do you know where that is?

Yes, up on the hill.

He framed that and had it all laid back there into a heap. You see, he would frame them all. So when he was ready, he said to the Searles, "Come on back up. I wanna raise the barn." He had what he called a sheer fall and he said he had a tackle block on and he could spring that around with rope and he hooked on that to pull up these here rafters. He said he had his pins and everything made, by God, and he put them up and shoved them in. They were all morticed, you know. They'd put them up and stick that pin in and you would think that they'd growed together. You couldn't even see a crack there. That's the way he framed the whole barn.

Well, a fellow was telling me when they framed the bridge, they would frame them in one spot, put them on a boat and float them down. They would take a bridge down and put another one up the same day.

That's right. That's right.

And he said he was a cracker-jacks carpenter, your father.

Oh, yes, he was.

I didn't know if he was a carpenter boss or if he had men working for him.

Oh, he was head boss.

That's what I thought.

Oh yeah. He was head boss. Oh yeah. Oh yeah. He done his work. And these here tools, like I said about coming up from #9 there that time, it was dark and this thunder and lightning shower came up and Boy! It was thundering and lightning so strong that the horse was trailing right along and then came a streak of lightning and the horse went right off in the canal. The wagon upset and his tools fell in the canal. He didn't bother with his tools or anything. He jumped right out and grabbed the horse and pulled the wagon right up and went home and ate. A few days after when the water cleared he could see where the spot was, he went back. He dove in there, got just about every tool out that was there. He was a good swimmer. Yessir, he'd go down. The auger was right down in the mud bank. He stepped right in.

Do you remember Dowling's Coal Yard?

Yessir. Mikie Dowling.

I understand he built that a few years before the canal was abandoned and that actually he never did get his money out of it because after 1915, there was no commercial boats that went through. The only boats that went through were repair boats and like that.

That is right.

And he tried to get coal and somehow or other, he lost out. He put a lot of money in that place.

Yeah.

Well, Art, if you have no more to add, I guess we'll terminate this interview. Thanks alot.

Dowling's Coal Yard, Stewartsville.

The Morris Canal and Banking Company had their own police force after the Lehigh Valley Railroad acquired the canal. Pistol permits from Phillipsburg were issued to Peter Kirkendall on January 9, 1882, Charles Segraves on September 3, 1883, and William Unangst on January 18, 1897. Unangst's permit also covered the Lehigh Valley Railroad.

Chapter Twenty

Dirty Ike's Work: Murder on the Canal

THE IRON ERA

No. #48 Volume XIX

Friday, October 31, 1890

DANIEL DAGAN, A BOATMAN, BRUTALLY BEATEN TO DEATH AND ISAAC HERDMAN ARRESTED FOR HIS MURDER

As if Morris county had not enough of homicides committed by her own people to smirch her reputation, non-residents must come within her borders to do deeds of fatal violence, the second case of that kind within a few years having occurred since our last issue.

William M. Dagan, of Broadway, Warren county, was an old canal boat captain along the line of the Morris Canal, and familiarly known. He was a man of muscular mould, about fifty years of age, and had a wife and ten children, some of the latter being grown. One of his boys, a lad of 14, boated with him this season, the remainder of the family being left at Broadway. Isaac Herdman was the captain of another boat who carried with him on his trips his wife and a boy who was not his son. He is a man of evil reputation and was known along the line of the canal by the sobriquet of "Dirty Ike." It is said that he of late was tied his boat up at Port Murray, Warren county, and calls that place his home. He is better known, however, as a resident of Newark, where the police are well acquainted with him, and from which place he has been sent to State prison.

These two men, both going East with their boats, arrived Friday evening at Plane No. 5, between Dover and Port Oram, where there is a stable belonging to the canal company in which boatmen may tie and feed their mules. Mr. Wm. Voorhees, the plane tender, says they were the only boats that put up there that night. Herdman arrived about 5 o'clock and put up his team, and Dagan got there between eight and nine o'clock. Mr. Voorhees did not see the men together on this evening.

Dagan's little son says that his father between five and six o'clock on Saturday morning left the boat and went to the stable to get his team. About this time Mr. Voorhees, the plane tender, saw a light at the stable and knew that one of the men was after his mules. He did not hear any words, or quarreling, and could not have heard them unless he were listening for them, as the roar of escaping water from the raceway between him and the stable is sufficient to drown ordinary sounds. About twenty minutes later he went out to the stable and saw Dagan lying on the ground near the door. Knowing that Dagan would sometimes get intoxicated he merely thought that he was in that condition and paid no further attention to him.

Meantime, Dagan's son, who says his father was perfectly sober when he went to the stable, had taken the boat across the plane and was waiting on the other side for his father to come with the mules. While thus waiting Herdman's boat came across the plane and the boy called to Mrs. Herdman, whom he saw on the boat, "Where's father?" To this the woman replied, "He and Dirty Ike had a fight and Ike done him up. He's up to the barn asleep." This frightened the boy, who at once went to the barn,

The stable in which William "Dan" Dagon was killed was located on the upper level of Plane 5 East, Port Oram near Dover.

where he saw his father lying unconscious. Mr. Voorhees also came out from his breakfast and found that the man was not drunk, as he supposed, but disabled. There was a cut over his right eye and a contused wound along the left side of the head, which had discolored the ear. There was no instrument or weapon lying anywhere near, but the leaves about the unconscious man were discolored with blood. He sent for Dr. Kice, of Port Oram, who came and washed the man's head and gave him such medical attention as he thought best. Mr. Voorhees, learning that Mr. H. C. Newkirk, of Dover, was acquainted with Dagan, sent for him. Mr. Newkirk at once went up and brought the unconscious man back with him in his wagon to Dover. Here Drs. Cook and Lindabury were summoned and did all in their power to revive Dagan, but without avail. Dagan's wife was also telegraphed for and came from Broadway in an afternoon train. The unfortunate man continued to grow worse until he died, at ten minutes after six o'clock on Saturday evening without having regained consciousness.

Coroner Hazen arrived the same evening, empaneled a jury, and directed Drs. Cook and Lindabury to make an autopsy. The physicians performed their work the same night and found that his skull which was of unusual thickness, had been fractured. Underneath the skull they found a large blood clot, causing compression of the brain, from which death resulted. In other respects they found him a strong muscular man with all the organs in a healthy condition. The blow which caused his death had been made with some blunt instrument, like that in use among boatmen and known as a "spreader." Through the efforts of Mr. Newkirk a purse was raised to meet the expense of preparing the body and sending it to his home, with his wife and son, early on Sunday morning.

Herdman's connection with the homicide was further established by his wife telling lock-tender Young as the boat passed through the first of the Dover locks that "Dirty Ike had got into a muss with Dan Dagan and had done him up."

Constable Roger Powell, of Mine Hill, was dispatched to Boonton on the 1:36 P.M. train Saturday to intercept

Constable Powell first tried to arrest Dirty Ike Herdman at Lock 12 East, Boonton.

A widewater at Boonton gave Dirty Ike an opportunity to jump from his boat to the berm side of the canal. Powell could not cross the canal until he came to a bridge, lock or plane.

Herdman's boat and effect his arrest. He returned on an evening train with the information that he had not been able to get his man. Several stories are told about the matter, but the one that Constable Powell tells to us is as follows: When he reached Boonton he went to the lock just west of the town and there met Herdman's boat. He tried to get to Herdman, but the latter, being acquainted with the lock through which the boat was passing, kept dodging out of the way. When the boat got through the lock Herdman was on it and the constable was on the bank. As he walked along Roger told him that he wanted him for assault and battery, and showed him his warrant. He tried to make him believe that the case was only trivial and gave him no intimation of Dagan's real condition. Herdman replied that he would go back with him as soon as he got his boat over the next plane, but when they reached a wide place in the canal Herdman used his boat pole to leap to the berm bank, on the opposite side, and ran away. Constable Powell, thinking he might return to the boat, intercepted it again when it reached Old Boonton, and searched it, but could not find his man.

That evening Ed Braxton, of this place, an old boatman, volunteered to catch Herdman if they would empower him to do so. Justice S.L. Williams deputized him and gave him a warrant, and Braxton, in company with Wm. Odell, started for Paterson on the ten o'clock train. At Little Falls they intercepted Herdman's boat, found him asleep in the cabin and dropped a pair of handcuffs on him. Herdman came with them willingly and they caught a coal train at Mead's Basin which brought them to Dover about six o'clock on Sunday morning.

They did not tell Herdman of Dagan's death, and he, supposing that he was wanted only for assault and battery, said he would fix it up with Dagan. He said that when they were loading their boats at Durham, on this trip, Dagan accused him of having hit him some fifteen years ago; that he pushed Dagan away from him and told him that he was mistaken; that he never did hit him, and that he never had any words with him. Herdman further told that when he went to the stable on Saturday morning he saw Dagan laying there. When he went back after his halter he was lying there still. He insisted that he knew nothing about how Dagan came to be there. He further stated that Dagan was drunk the night before, and that his breath on Saturday morning stunk bad enough to knock a man down. How he knew that he did not say, and his story conflicts with that of the Dagan boy, who says his father was entirely sober.

While in Justice Williams' office, waiting to be conveyed to Morristown, Herdman had a conversation with Mr. Voorhees, the plane tender. He asked very anxiously if Mr. Voorhees had heard any words or quarreling between him and Dagan, saying he was afraid somebody might tell falsehoods about the matter. He then told Mr. Voorhees that on Saturday morning he met Dagan coming into the stable as he was coming out with his mules. He said Dagan came up to him and he pushed him away, but he would swear that he did not have a stretcher and that he did not touch him. This was a different story from the one he told to the effect that he first saw Dagan lying on the ground.

A few slight scratches were noticed on the left side of Herdman's face after his arrest. He was driven to Morristown by Jas. Gardner, in charge of Constable John F. Wood, and lodged in the county jail. A large crowd of people was on the street to see him taken away. The Newark papers say that "Dirty Ike" has been a notorious character and that he once kept a low dive in that city, which was closed by the police. He was sent to State prison from Newark. He was also at one time in the Morris county jail, having once been arrested on his boat at Rockaway by Detective Wm. B. Goodale, for having threatened the life of Mr. R. K. Stickle, of that place.

Later we learn that Herdman has served two terms in State Prison, once for burglary and again for bigamy, he having married a girl named Delaney, who lived in Newark, while he had a wife living. He was always very hasty with a weapon, and many an unfortunate was picked up in front of his dive with a broken head, after having been clubbed and thrown out of his place.

The funeral of Dagan was held at Broadway on Tuesday, and being an old veteran, he was buried with military honors by the G.A.R. Of his ten children five are under sixteen years of age.

Herdman denied positively Constable Powell's story of his attempt to escape. He says the constable showed him his warrant and that he told the officer he would go with him willingly after getting his boat over the plane, as his wife could not get it over. After he had got the boat over the plane he says the constable was nowhere to be seen and that he then went on his way unmolested.

THE CORONER'S INQUEST

was convened at 10 o'clock on Wednesday morning in Justice Williams' office by Coroner Jas. C. Hazen, of Morristown. The following were the jurors: Chas. H. Munson, Andrew F. Rice, Sylvester Dickerson, Jas. H. Brown, Jas. P. Kelley and Andrew J. Messenger. Coroner Hazen instructed them as to their duties and the evidence was commenced.

Dr. H. W. Kice, of Port Oram, related that he was called Saturday morning to the Dover plane by Dagan's son. He found Dagan in the stable, lying on his back, in a state of deep coma, from which he could not be roused. He found a little cut above the right eye, and noticed redness about the left ear, which later turned to black and blue. He considered it a hopeless case and advised his removal. Some around there were talking that the man had had a fight and they supposed Dagan had been struck. He did not know who brought the blankets with which the man was covered when he first saw him. He thought the case hopeless and did not expect the man would recover.

Wm. C. Voorhees, the plane tender at Plane No. 5, testified that he owned the barn. He saw the mules of Dagan's boat stabled there. His boy put them there on Friday night. Herdman had a horse and mule and they were put in there at 5 o'clock. He saw Herdman that night, but not Dagan. Herdman went right back to his boat at the head of the plane. Dagan's boat was at the foot of the plane. He supposed he saw Dagan going to stable with a light about five o'clock in the morning. When he first saw him to know him he was lying on his face in the stable. This was 10 or 15 minutes after. He did not see Herdman till about six o'clock on his boat going down the plane. He first tried to rouse Dagan, but could not do it, and thought he had been drinking and might wake up himself. He was snoring like a man in a deep, sound sleep. He went back to his breakfast and returned to the stable and tried to rouse him, but could not. He was then in a sitting position. This was before Herdman left with his boat. He did not see anybody else anywhere. When witness saw Herdman he expressed wonder as to what ailed Dagan, but does not remember what Herdman said, except that he was anxious to get Dagan up and on his boat. After this the boy came to the stable to see about his father. He asked the boy if his father had been drinking or had liquor on his boat. The boy said he had not, and had not been the day before. The witness covered Dagan with his own blanket. He had not heard any boatman speak of any dispute between Dagan and Herdman. He had heard they had a dispute at Durham, while unloading on his trip. Herdman told him this after his arrest on Sunday morning. Herdman said he did not know how it happened. He met Dagan coming in the stable as he was coming out and pushed him away, but knew nothing more. Herdman's horses were removed. He did not think the man could set up himself. His eye seemed to be plastered with something, like tobacco. He was sitting against the partition of the stall, with his head hanging forward on his breast. He had evidently been set up by some person.

Dr. W. L. Lindabury testified to being called to attend Dagan after he was brought to Dover. It was about 1:30 in the afternoon. He lay on a cot in a comatose state. All the marks about him were a cut over the right eye and a bruise over his ear, extending back of the ear. It was a bruise. The skin was not cut. There were a few scratches on his face. His impression was that he was sick from injuries and that he would not recover. He was called to make an autopsy Saturday night. They found a slight fracture of the skull and a blood clot about the size of your hand. It was compression of the brain. He told how a further examination showed the stomach, heart and other organs were all-right. Death was undoubtedly caused by compression of the brain, due to the blood clot. There was a clot of blood between the skin and the skull, showing the blow had been made by a club or some other blunt instrument. He hardly thought such a wound could have been made by falling down. The bruise was about horizontal, extending straight across the head. If a person of the same size should strike another the wound would be a little higher behind, as it was in this case. The blow, he thought, would have come from in front. In his opinion he judged the injury had not been received from a fall; it might possibly have resulted from a kick, but was more likely to have been done by a club or billy. Such a wound would hardly be made by a kick of a horse or mule. The blow might have been made by a sand bag or something of that nature. The blow might have been made by a hard club and not break the skin.

Henry K. Youngs, who tends the first of the Dover locks, known as No. 4, said he knew Dagan. He saw him on Saturday morning in the stable at the plane. He had heard he had been hurt, and went up to see what he could do for him. He saw a boat pass through his lock on Saturday morning before he saw Dagan in the barn. It was Isaac Herdman's boat. It was about a quarter to 7. Herdman told him Dagan's boat laid at the foot of the plane and witness asked him if Dagan was coming on, and he shook his head as if he didn't know. Witness's son told him that Dagan's son had said to him his father and Ike had had a fight and Ike had given him a good licking. A Mrs. McClay had told him that Mrs. Herdman told her that Ike had killed Dagan.

Joseph F. Dagan, the son of the dead man, said he was going on 14. There are ten of the children. He was with his father on the boat last Friday night and slept with him. I put the horses away. He was at the boat all the time. Father had not been drinking Friday night. I saw him leave the boat a little after 5, to go for the mules. He had a lantern in his hand. He said his head ached. The boy saw his father go in the stable and saw the lanterns hanging there. He was about to go up to see what was the matter when Herdman's boat came down. He asked Herdman's boy what was the matter with his father that he didn't come, and he said he didn't know. Herdman and his wife talked together and then Mrs. Herdman came to him and asked him if his father was drunk. He said no. Mrs. Herdman then told him that his father and Ike had had a fight, that Ike had given him a good licking, that he laid up there in the stable, and that she didn't know what had been done. Herdman asked him if his father had any rum in the boat. The boy then told how his father was lying on his back when he first saw him in the stable. The boy said Herdman unloaded ahead of them at Durham. He did not know that they ever had a quarrel there or anywhere else. His father was 52 years old. He saw Herdman go to the stable. After Herdman came away from the barn, he went back and got the lantern and brought it to the Dagan boat.

Thomas Swan testified that he was brakeman at Dover plane and was there Friday night and Saturday morning. On Friday night Dagan was well and appeared to be perfectly sober. He testified to seeing him in the stable the next morning. Herdman told me Dagan was asleep there. Herdman also told him that Dagan was going to fight or wanted to lick him, or something of that sort. Witness asked Herdman if Dagan was drunk and Herdman said he was not. He said Dagan was up to the barn asleep, but did not say anything. The witness said that Herdman came to the stable after he and the plane tender were there. Herdman tried to wake Dagan up. Herdman told witness that Dagan came to him and wanted to lick him. The cut over Dagan's eye looked as if it was plastered up with tobacco. Herdman, when he came there, raised Dagan up like, and called to him, "Billy, wake up." He also wanted to move Dagan to his boat.

Dr. R. I. Cook testified to having been called to see Dagan in Dover and to making the autopsy with Dr. Lindabury. His testimony corroborated that of Dr. Lindabury. He was a man apparently in good health in every respect. In his opinion the cause of death was compression of the brain, caused by the blood clot. He thought the injury must have been done by some instrument with a blunt edge. It might have been done by a sand bag or a club covered with something. In his opinion it was not made by falling or the kick of a horse or mule. No evidence of liquor was found in Dagan's stomach.

Henry C. Newkirk told of how he first inquired on Saturday of plane tender Voorhees and brakeman Swan as to the cause of

Dagan's death. No one could tell what was the matter. He also told how he telephoned to Broadway to Dagan's wife and a friend at Broadway, and of how he removed Dagan to his place in Dover. Witness related how a boy coming up the canal on a boat soon after told him how Herdman had said to him that he had "had a fight this morning with Dagan and had done him up."

Wm. Kiefer, canal foreman, said he heard Swan say that Dagan's boy had told him that Mrs. Herdman said that Ike had given Dagan a flogging.

Roger Powell, constable, told of his ineffectual attempt to arrest Herdman, as given elsewhere.

William Fahrell, lock tender at Boonton, said Herdman did not pass there on Saturday, although his boat went through at 3:15. He heard Mrs. Herdman talking with Bill Erwin, the captain of another boat, and heard her tell him that Ike had

Plane 7 East at Boonton was where Constable Powell had hoped to capture Herdman.

done Dagan up, and ought to have done it before.

Edward Reilly, locktender in Dover, told of Dagan's boat passing through his lock. One boatman told him this trouble began in Newark 17 years ago and that Dagan was struck in a rumpus there. The same man said Dagan tried to get into a fight with Herdman at Durham. Dagan went into Herdman's place in Newark for a boy when the original row began. The witness noticed some scratches on Herdman's face when he went through and thought likely he had been in some trouble.

Jas. Bell, the Marshal, saw Herdman on Sunday morning in Braxton's house. He said that if Dagan was living he would say he didn't strike him. He said as he was taking his team out of the stable Dagan came there and he told him to go away. He said he and Dagan were always the best of friends.

Edward Braxton told how he arrested Herdman. As he entered the boat he heard Mrs. Herdman say, "Ike, I told you

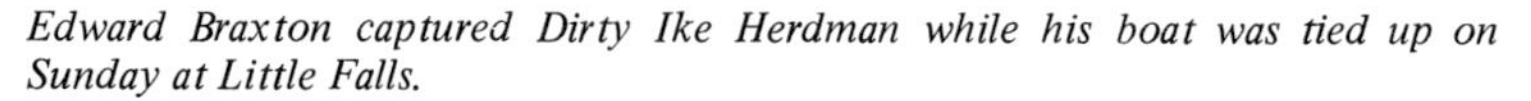

Edward Braxton captured Dirty Ike Herdman while his boat was tied up on Sunday at Little Falls.

there would be trouble about this stable." When he told Herdman that Dagan was dead Mrs. Herdman said, "For God's sake, Ike, you've killed Billy Dagan." Before he knew Dagan was dead Herdman said that he thought he would get out of the county and when Dagan came down would settle it with him by giving him five or six dollars. He said that if they hang him for it he was innocent.

Wm. Odell, who accompanied Braxton, corroborated the testimony of Braxton, he having accompanied the latter in making the arrest. Herdman said that as he came out of the stable Dagan caught him in the face and he pushed him away. He said that Dagan was intoxicated. He said as we came along, "God knows I never hit him." We did not ask him any questions.

John F. Wood, constable, told that while taking Herdman to jail, the latter said, in reply to a question as to what was going on there, "We had a racket." The constable asked him what he hit Dagan with and he made no reply.

At this juncture the inquest was adjourned to Saturday morning, at 10 o'clock.

SEQUEL:

A trial was held. Isaac Herdman was convicted of manslaughter and sentenced to eight years in prison.

NOTE:

Daniel Dagan could have been a nick-name. The last name should be Dagon instead of Dagan. He's now buried at Phillipsburg Cemetery. William Dagon's four sons–Joe, Beecher, John, and Charlie–followed the canal and later they all had boats of their own.

FLORENCE VAN HORN

Born at Washington in November 1890. Daughter of canalboat Captain James Campbell (Camel). Now living at Washington.

Chapter Twenty One

Florence Van Horn

"Well, I just fixed something to eat. We didn't have time to get much through the week. They had bacon and ham and eggs through the week–something quick. But then on Sunday you always try to make a stew or make a pudding . . . "

This is a taped interview between Mrs. Florence VanHorne and Mr. James Lee, taped September 15, 1975.

Now, Mrs. Van Horn, how old are you?
85.

You are 85. And you will be 86 in what month?
In November.

What was your connection with the Morris Canal?
Well, I liked to be on the boat and I liked to go from one end of the canal to the other; from Phillipsburg to Jersey City Basin.

And your father was a captain?
Yes.

And what was his name?
James Campbell. I think he had a middle name too. James Elliot Campbell.*

*Editor's note: In the poem *Famous Tiller Sharks,* James Campbell's name was spelled "Jim Camel"

He had a little boy, a young man driver to boat with him.

And you boated along with him? How old were you?
About 12.

Twelve years old. What did you have to do when you went along with him?
I used to wash the dishes and cook. You had a little stove in the cabin and it had a big furnace near the hinges of the boat.

And did you walk the mules sometimes too?

Yeah. I drived sometimes for exercise. Then when I got tired I got on the mule's back.

Oh, you would? What were your mules' names, do you remember?

Katie and Jenny. Jenny was a kicker. I couldn't touch her. But Katie, I could climb up on her back and ride for awhile.

Did you walk barefooted?

Yeah, I would walk in my bare feet.

Did you ever worry about walking barefooted?

No. We had a nice towpath.

Was there anything you were afraid of stepping on maybe?

I was afraid of different animals—outside of snakes.

That is what I mean.

But snakes—generally I seen them in the water swimmin'. They didn't stay on the towpath. They might go across it, but they didn't stay. They was in the water.

James E. Campbell (Camel), Mrs. Van Horn's father, was one of several black canalboat captains on the Morris Canal.

Now, how many black people were on the canal that you know? Quite a few in this area?

There was quite a few, but I can't remember their names. Mary Taylor and Joe Taylor, they lived here. And Mr. Haines lived here. There were a lot of men that lived here—all the way up to the end of the street, but I can't remember them now.

There is a poem in the back of this book and it says, "Happy old Jim Camel, Jim Haines is happy still". Was your father happy?

Yessir! He was a good man.

He was generally pretty happy?

Yes, he was. He would do favors for anybody and help anybody. He was that type of person—very good. White or colored, he would do you a favor. Yessir. He was a nice man and he worked in the church. He worked, after he worked, can you remember when they abandoned this canal?

Yes.

Was it 99 years or 90 years?

It lasted about 99 years. They abandoned it in 1924.

Yes.

It didn't last quite 100 years as far as the charter was concerned.

And then after he got off the canal, he worked for the Canal Company for awhile after he stopped boating. And then when they cleaned up everything and got rid of the canal, he worked in Washington for the Water Company until he died. He always followed the water.

How was life in those days compared to now? Was it easy-going?

Well.

Did you have a pretty good life, did you?

He made enough money on the boat to keep us all winter.

Did he have a job in the winter?

No, not every winter. He didn't work.

But when he did work, what kind of work did he do?

He worked at Terra Cotta works in Port Murray. Isn't it?

Yes, Port Murray. There's one up there.

It was the brick house somewhere up there he worked.

How were the boys and girls in those days compared to today?

Wonderful, wonderful. I had nice colored boys and nice white boys too. 'Cause I went to Taylor Street School and they were very nice to me. There was only just a few nasty people that would call you "nigger" and chase you and try to fight you. But another bunch of white folks would take up for me. Walk me from Taylor Street School down to the corner to see that I got home. I had lots of white friends in my day. Girls used to come and play jacks with me on my porch and I used to go to their house and play jacks.

What other games did you play on the canal when you were on the canal?

Oh, on the canal?

Yes, when you were on the boat.

When we went to the, if you go to the end and, let me see now . . . At closing time for the boats when you put your mules in the company stables, well, generally at those locks or planes most of them had a little store where they sold soda and candy and stuff for the boatmen. Whenever we got there around the stores I was right with the gang. We was all out there. They had jew's harps, mouth organs, and bones.

Bones?

That was the music, you know.

Oh yeah.

And they would get out there by the store and play that music and all us girls would dance.

Do you remember any songs that you might have sung?

No. I can't remember the songs.

Do you remember any songs except those that were popular songs of the day?

No. They played the music and I remember that we danced and they would buy us soda and all kinds of stuff from the store.

But you don't remember any canal songs?

No, I can't.

What about poems?

No. See we would only meet at the store, the gang on the boats. Then we all would go one way or another. We didn't stay together, only when we met at the stores or somewheres. But we knowed everybody practically.

Do you remember that one, "You Rusty Old Canaler, you'll never get rich?"

No.

You don't remember that one either?

No.

No? No poems either about the canal?

No.

How about on Sunday? Did you sing a lot on Sunday when you got together?

No. Sometimes we would be to a lock where there was only a company stable and we would be in the woods on both sides of us and my father, he used to go fishing a little bit. But I was against fishing!

Why?

I was teached not to; teached from young in Sunday school not to fish on Sunday. One day I went fishing one time and they told me I caught the devil. The devil would say, "Take me home, cook me, and eat me up." And that they told me when I was young and I was against fishing.

And that kept you from fishing? You didn't do too much then on Sunday?

You see when I first went up to these woods close to the canal near the towpath and the berm, well, here would be the woods and there would be the woods and here would be this lock. Well, I used to take a pin and go out and fish and they told me about me catching the devil and the Sunday school

Feedboxes were located at the "hinges," the middle of the boat where the two sections joined. The oats section was used to store ham and other meats.

Many inclined planes had stores closeby where the drivers or children on board could buy sweets or baked goods. Location: Plane 11 East, Bloomfield.

teached me, I was teached in Sunday school to remember the Sabbath Day and keep it Holy and not to fish and not to work. I didn't mess that. I kept it all the way on the canal. I kept my commandments with my church.

What did you do on Sunday then?

Well, I just fixed something to eat. We didn't have time to get much through the week. They had bacon and ham and eggs through the week–something quick. But then on Sunday, you always try to make a stew or make pudding or something, you know, different to cook. And you didn't get up so early either.

That was a day of rest. How about salt mackerel? Was that a favorite dish?

Yes. We used to buy them by the pail. They called them kits. A kit of mackerel and you would keep them down by the hinges. They had a place for them in the cabin.

Where would you keep your ham?

In with the oats. They had two of them . . . what would you call them? Feed Chests. And they had them full of oats and one for feed. And down in the oats the eggs were kept.

What else might you keep there?

Well, they kept the hams all wrapped up there.

Yes, they were salt hams probably.

Yes. Bacon and like that.

Would you keep anything else there besides the meat and the eggs?

No, that was about it.

Did you ever get fresh milk?

Once in awhile when we got into a town like Bloomfield or Paterson where there was stores and they sold those things and maybe a Baker Shop.

When you went along to various towns, did you ever go to churches there?

I did go at Bloomfield, New Jersey. At the head of the plane we had to stay from Saturday over to Sunday and then the minister came over to all the boats that laid there. And he came over to get all the children for Sunday school. And we were on the boat, my sister Loretta and myself and for us girls we had excuses because we had no Sunday clothes and no shoes. But he nagged us to come and just kept at us. My father said, "Go". So we went to the church at Bloomfield with our bare feet and gingham dresses–just everyday dresses, gingham dresses. We went to church, to Sunday school and I enjoyed it because they used us very nice.

Would you get little cards when you went to Sunday school?

No, I can't remember.

What else do you remember? Do you remember any good stories from the canal or any tragedies that happened on the canal?

I can't remember any. Oh, oh, oh! Only once in awhile this big fat lady and little skinny man and they had lots of children and every once awhile him and her got in a fight and she would beat him up. They had a lot of children, a lot of babies on the boat. And they had the babies fastened to the cabin

Wives tied their children to the towing post or deck to prevent them from falling into the canal.

with the chains so they wouldn't fall. And he would get in a fight with his wife and she would beat him up. And sometimes he would have to jump in the water!

To get away from her?

To get away from her.

She was really the captain then.

Yeah. That is the only tragedy I can remember. That was along when we came up from Paterson.

Did you ever fall in the canal or did any of your family ever fall in the canal and almost drown?

Oh, for goodness sake. O Lord, I would have been the one to drown if I fell in because I can't swim.

How about your sisters or brothers? Did they ever fall in the canal?

Oh . . . my brothers died. They never got on the canal. Beautiful brothers I had. We had a boy that worked. We called him a boat hand. You know, to drive the mules and take care of them like that. But we didn't have no boys, just my sister. She was like a boy. We called her "Tomboy". She liked to climb and drive the mules and take them around better than me because I was kind of slow about that.

You told me something that happened at Saxton Falls one time when your sister almost drowned there. She was running or something. What happened there?

Yes, that was Guinea Hollow Dam.

What happened there?

Well, it was raining and Mother told us not to run. We was on the hinges. That is the stern part of the boat. And the bow was on the other side. You can open up the bow with the hinges and you can put them back together again.

The boats were in two pieces then?

Yes.

You could couple them up again? The front section and the back section?

Yes, you could do that. We were there and the boats were

together. We were coming along fine, but a rain came up and made the—what do you call the running board?—it made it slippery.

Yes, it made the decks slippery.

She said, "Don't run" and my sister didn't listen and didn't mind and she run, she run ahead of me and plunged into the water. And scared me to death. And then I got up there and my pa jumped in and he could not manage with a coat because it was a big overcoat he had on. And he had to go to the towpath and get the overcoat off. No, he didn't go to the towpath. He went to the boat. And he got up on the boat to take the coat off. He got the pike pole. I called it a pike pole. A great big long pole—it has a hook on it. And that is what he had to do. Now, he kept hollering, "Hanna, Hanna! Steer the boat! Hanna, steer the boat!" and she did make the lock. And he stayed back there and got her with the pike pole there from her dress. He could have hurt her with that pole, but he managed to get her with the dress and pull her out on

The pike or boat pole was a versatile instrument. Many lives were saved because of its availability.

the towpath. Then the man, the Lock man, Mr. Burd, was it? I can't remember too good names. He came on down and helped roll her over the pail and give her whiskey and got her together so they could put her back on the boat.

In the book here in the back . . .

I only remember the Stackhouses.

Do you remember Stackhouses? "Happy old Jim Camel", of course, that was your father. And Jim Haines, that was your neighbor. Your father and Jim Haines, were they good friends?

Oh, yes. Mr. Haines would have a barn out back of his side and my father had a great big barn.

The barn for the mules? He kept the mules there?

Yes, my father had a place for a haymow. An awful big barn he had.

How about the Healys and the Tuckers?

Well, I only know the Tuckers from Newark, but I don't know about them boating. But they could have boated. A lot of them lived in Newark that boated. That is why they didn't live in Washington.

But then there is Bill and Amos Stackhouse?

They lived in Newark too.

And then there was the Jewel boys?

They all lived in Newark. I think there was a Jewel that was a locktender.

Okay. And there were some white boys that you knew too? The Dagon brothers maybe? The Gundermans?

The Gundermans and the Peers. And there was another family I used to know. Oh, it won't come to me now.

Did everyone get along good on the canal? The white captains and the black captains?

As far as I know they did. There was never no trouble.

There was never no trouble?

No, 'cause I know when a mule or a horse would die there on one of the boatmens, why, then the other boatmens would try to get together and get him a substitute right away and try to help him. Like you do when your car is down and then you go get another car. They helped one another. They were a good kind. They were poor people that had the boats, but they were good to one another. They would do that–help them. Anyone sick on the boats, they helped. If anyone was sick, they would see that you had a doctor. Now one time my pop went up to the stables to put the mules up and told me to stay home and stay in the cabin. I didn't stay in the cabin. I went to the hinges and was making him coffee. In them days we had coffee pots and I went to the hinges and was fixing coffee in the coffee pot and I scalded this leg. The scar is there yet. I was all alone and scalded my leg so I went to the side of the hinges and the cleats. You know, what they kept the rope around. I got ahold of the cleates and got my body down in the water. I could have been drowned. I put my scalded foot down in the water. I didn't know what to do. Then after I got down in the water and cooled my foot off, I went back to the cabin and my pop was awful mad at me for I didn't stay in the cabin. For I tried to make coffee for him when he came back, but I made a mistake in doing that. Of course, I had an awful scald on my leg and he had to go to the drugstore and I was about two months getting better. I didn't have no doctor, just drugstore doctors.

Would you say they were the good old days?

I think so. I was happy. I didn't have no troubles like I am having now. No, I didn't have no bills to pay and I didn't have no nasty people around like there is now. Everything must be locked up. You can't have nothing out. It is miserable for me now.

Yeah.

Course, I am alone and they did come in and climbed my upstairs window and robbed me. They took my clothes, money, and my best everything. Pictures and everything.

But that didn't happen years ago?

Years ago you didn't have to nail your windows down. People

used to have screens in the windows. All summer 'round you would see people with screens and some poor people would nail screening over their windows. They were nice in my day. I had no trouble.

Well, Mrs. Van Horn, I think we have a good tape here and I want to tell you how happy I am that I got to talk with you this afternoon. Thank you.

Members of the maintenance crew begin to dismantle an inclined plane car. An era in New Jersey's transportation history was over.

OLIVER WARMAN and SAMUEL WARMAN

Oliver Warman, canalboat captain: born at Broadway on June 25, 1877 and died at Phillipsburg in 1963. Son of Samuel Warman, also a canalboat captain.

Chapter Twenty Two

The Morris Canal in 1894 by Oliver Warman

As told to Mr. James Lee by Mr. Oliver Warman in 1960

The Morris Canal was a distance of 102 miles from Phillipsburg, New Jersey to Jersey City, New Jersey. Water was supplied by Lake Hopatcong, the highest point along the way. Boats going East had to go against the current to the Lake, and with the current to Jersey City. To obtain the highest points there were 11 incline planes to the Lake and 12 on the Eastern slope to Jersey City and also 29 locks in the 102 miles.

Boats at that time were owned by the Canal Company and teams of mules by the Captains. All loads going East were of hard coal for coalyards along the way; mostly Paterson, Bloomfield, Newark, and Jersey City; and also up the Passaic River to Lyndhurst, Bellville, and Passaic City; also up the Hackensack River to Hackensack City and Englewood, New Jersey. Loads ran from 65 to 80 ton and the amount of freight payed varied as to the distance taking the highest 40.8 tons to Jersey City. (The average trips for a season was 22, which started on April the 1st and ended when cold weather froze up the canal in the fall.)

The Captains lived along the canal in many towns, but had to leave their boats where they froze up and had to get to their homes the best they could. I was caught ten miles this side of Paterson and lived at Washington, New Jersey. I had to come home with the mules from there with ice on the canal that you could walk on. I left there at 9:00 a.m. and got home at 10:00 p.m., a good long cold walk. We had to get out in the morning at 4:30 as they started to operate the locks and planes at 5:00 a.m. and continued until 9:00 p.m., and on Saturdays until 10:00. Sundays we did not work, but on Mondays we were on the go at 5:00 a.m.

Most Captains had two mules; some had one, but mostly mules, not many horses. Mules were tougher and could stand the long drag better, but they had to be fed good—4 quarts of oats at 7:00 a.m. and at 11:00 a.m. and at 4:00 p.m. and what was known as a "cut mess" consisting of chopped corn and oats. They got that when their day was in sometimes at 9:00 p.m. and very often later.

Some of the towns we passed through after leaving Phillipsburg: Hard Port, Stewartsville, New Village, Broadway, Brass Castle, Washington, Port Colden, Port Murray, Rockport, Hackettstown, Waterloo, Stanhope, Port Morris, Shippen Port, Lake Hopatcong, Drakesville, Kenville, Port Oram, Dover, Rockaway, Denville, Powerville, Boonton, Montville, Beavertown, Lincoln Park, Singac, Mountainview, Little Falls, Paterson, Bloomfield, So Ho, Forest Hills, Newark, Greenville, Communipaw, and Jersey City.

Sheds were built over the lock mechanism to guard against bad weather and to give shelter to the lock tenders. The towpath was ideal for a Sunday outing.

Chapter Twenty Three

Notes from the Newspaper

NEWARK SUNDAY CALL
JULY 26, 1891

To The Editor of the Sunday Call:

Much has been written about the low water in this lake, but much of it is purely nonsensical, and it is undoubtedly the result of jealousy and envy.

The original sale to the Lehigh Valley Railroad Company was supposed to be part of a large and comprehensive system of supplying water through the canal to Newark, Jersey City, New York, etc. The Lehigh Company has put in a dam and raised the original surface of water three feet above the ordinary level last year as is shown all around the lake on the rocks, etc. This water was to be held in reserve and used when wanted in dry times. The increased rise covered acres of low lands, stumps and rocks, and added very largely to the beauty of the lake, its value for shore frontage, and enlarged its area by a thousand acres, more or less, which to the riparian owners became a vested right.

The Hopatcong Steamboat Company built two side-wheel steamboats, the HOPATCONG with a capacity of 150 passengers and drawing 22 inches of water, and a launch of a smaller size, and to avoid the charges of lockage in the canal the Steamboat Company employed a dredger and excavator to remove the stumps and cut a new channel one and a half miles long from near the Hopatcong railroad depot to deep water in the lake and this channel has been used by the MUSCONETCONG carrying the passengers and freight between the depot landing transferred to and from the larger HOPATCONG at deep water during the completion of the rest of the channel.

To cripple the new line the water has been drawn down, but it has failed of its purpose, as the new boats draw less water than the propeller launches.

Boats of the Hopatcong Steamboat Company had their own dock and used a special channel to enter Lake Hopatcong.

A boat belonging to the Lake Hopatcong Steamboat Company, the line which used the canal.

The opposition thus created has been the cause of lowering the water of the lake, exposing thousands of stumps and rocks, endangering navigation, and despoiling the beauty of the lake in many places by exposing stumps and rocks referred to, and in some cases endangering the health of citizens along the bayous.

To prevent all this proceedings are about to be instituted, and Cortland Parker will probably be selected as the champion of the people to restrain the changes of line and surface. Nice questions of law are involved and the whole of New Jersey are interested in maintaining the great beauty of Hopatcong, regardless of a squabble between rival lines of steamboats.

J.L.D.

NEWARK SUNDAY CALL–JULY 26, 1891
LAKE HOPATCONG

The Steamboat Quarrel and Low Water

When I visited the Lake last week I brought my rubber boots with me, expecting that I should have to use them in walking across the damp bottom of what was once beautiful Lake Hopatcong. The exaggerated reports of the rapid decrease in the water gave the idea that by the time one reached Hopatcong the lake would have entirely disappeared. In this visitors were happily disappointed. The total fall of the water is about three and a half feet. This is particularly noticeable at the southern extremity of the lake at the steamboat landing. The cause of this is the Morris Canal Company. If you ask why the water of the lake has been lowered the agents and sympathizers of the canal company will tell you the past dry season has compelled the company to draw upon the lake to an unusual extent to supply the canal. As it happens, this has not been an unusually dry season, and people who ought to know say the canal company has no use for the extra drain. The true inwardness of the matter is believed to be the competition between the two steamboat companies which carry passengers to all parts of the lake. One is known as the Lake Hopatcong Steamboat Company; the other is the Hopatcong Steamboat Company. The former has several steam launches which takes passengers from the railroad station landing, charging each passenger 50¢ for each trip. The latter company has two large and nicely-equipped steamboats, and passengers are charged only 25¢ each to be taken to their destination while return tickets are sold for 40¢. The former company takes the passengers through the canal. The boats are small, overcrowded and uncomfortable. The regular visitors, as a rule, patronize the Hopatcong Steamboat Company. To reach this company's landing, it is necessary to walk over the bridge above the railroad tracks. It is but a minute's walk, but as the boats and their dock is not in sight the strangers often patronize the high-priced boats. The launches which use the canal are a source of revenue to the canal company; the other boats are not. The canal company began drawing the water off the lake and left the rival company's boats high and dry. The latter, not to be thwarted in this manner, got a dredger and began the work of making a deeper channel. This has been successfully accomplished, and as long as the canal company continues drawing water off the lake the dredger will be kept busy widening the channel.

FREIGHT HITS RAISED DRAW

Train Crashed Into Bridge and Tender Falls on Canal Boat

CREW HAS A NARROW ESCAPE

Special Dispatch to the EVENING NEWS

Dover, June 13, 1905—Slippery tracks, a down grade and defective air brakes were jointly responsible for an accident which occurred yesterday on the High Bridge branch of the New Jersey Central Railroad, and resulted in a badly damaged locomotive, tender and a high-side gondola, the total demolition of a canal drawbridge and the sinking of the after part of a canal boat. Incidentally, Skipper George Meyer's daughter Katherine, four years old, who was at the time asleep in the cabin of the boat, narrowly escaped drowning, after also having had a narrow escape from being crushed to death by the trucks of the tender, which dropped on the boat, a peril which her mother, who was at the tiller, had shared.

The little girl was so frightened by the accident, it is said, that she was struck speechless and up to today had not recovered the use of her voice.

BRIDGE WAS RAISED

The scene of the accident was at the guard lock immediately east of the Dover Station, over which there was a lift drawbridge. Meyer's boat, laden with coal, had just entered the lock, and the bridge was consequently raised with signals, it is said, set against trains, when a coal train, consisting of a "camelback" engine, tender and nine loaded gondolas, came down the road at a fair rate of speed. Engineer Henry Schafer, of Mauch Chunk, applied the air brakes as soon as he caught sight of the signals, and when he realized the failure of the brakes to hold the train, he tooted sharply for "down brakes". The train kept sliding

One June 12, 1905, a collision occurred between the Central Railroad of New Jersey engine #410 and a canalboat at the Guard Lock, Lock 7 East, in Dover.

along, however, with no perceptible slacking of speed, and a warning blast from the whistle made his fireman, M. C. Holman, also of Mauch Chunk, leap to the ground, only a second or two before the engine crashed full tilt against the raised draw, with the result that it carried the entire substructure, heavy steel girders, ties, rails and all, along with it.

About half a car's length the other side of the lock the engine and draw came to a stop, and Schafer descended from the cab uninjured. In place of the drawbridge the lock was spanned by the first gondola, which was broken in the middle, while between the gondola and the engine the body of the tender stood on end, both front and rear tracks of the tender having dropped down on the canal boat in transit.

CABIN FILLS WITH WATER

It was Mrs. Meyer's first impulse, when the trucks came crashing down immediately in front of her, to leap from the deck to the retaining wall of the lock, but remembering her child she tried to get into the cabin, which was rapidly filling with water. The trucks barred entrance and she reached down to the bunk on which the girl lay, and to the top of which the water had already risen, only to find that the child was held fast by some wreckage, the trucks having broken partly through the roof of the cabin and the after deck. With the aid of her husband, Mrs. Meyer rescued the child with only slight injuries. A wrecking gang worked all night and today has a temporary bridge built across the lock and the tracks restored. As the makeshift bridge cannot be raised, the canal will remain impassable at that point until a new drawbridge is placed in position.

MORRIS CANAL LORE
Ghost of a Chance Fails

Ghosts on the Morris Canal? Well, according to Douglas Williams of Montclair, New Jersey, there was a gambler from

Paterson who won a large sum of money playing cards in the late 1880's. Afraid that the losers would follow him and try to steal his winnings, the gambler left by the Morris canal. He was never seen or heard from after that night.

One canal boat, the "Lager Bier," used to tie up in the Brookdale section of Bloomfield near a huge old oak. After the gambler's disappearance, it was rumored that some of the boat's crew had murdered him for his money and buried him under an old oak tree.

To add more fuel to the gossip, residents began to claim that they had seen a ghost sitting in the branches of the oak. Soon after the ghost rumors started, the skipper and two crewmen of the "Lager Bier" quit canaling.

Today only a sharp eye can spot the bit of canal that remains in an indentation along the Garden State Parkway. The Ghost Tree has also been removed, but its tale lingers in the memories of the prople who lived along the Morris Canal.

THE EVENING JOURNAL, JERSEY CITY
TUESDAY, APRIL 14, 1891

WENT DOWN WITH THE BRIDGE

A LEHIGH VALLEY LOCOMOTIVE DUMPED INTO THE CANAL

There was an animated scene at the Lehigh Valley Railroad bridge over the Morris Canal, near Pacific Avenue, this morning. Gangs of mechanics were working on a run, and a constantly lenthening line of canal boats were forming on each side of the bridge. The bridge was the focal point of interest. A bob-tail locomotive used for drilling trains around the yard had been run on the bridge when the bridge had not been properly closed. The bridge and the locomotive were dumped into the canal together.

A wrecking gang was summoned, and all night the flicker of lanterns showed where men were at work. The engine was hoisted out of its bath and a new bridge took the place of the old one. The canal boats stood in a row, the teams still attached, waiting for an opportunity to pass.

NEWARK DAILY ADVERTISER–JUNE 1, 1878

STEAM ON THE CANAL

Mr. William Baxter's steam canal boat recently completed at Washington, Warren County, made her first trip this week arriving in Newark at 5:00 p.m. yesterday. Today at noon she proceeded to Jersey City to discharge her cargo.

Complaint is made by boatmen of the Morris Canal that they experience much difficulty in navigating this season account of the long grass which has grown to a considerable extent in some levels.

The patent model submitted by William Baxter and William Baxter, Jr., to the United States Patent Office in 1874 for an "Improvement in Steam Canal-Boats."

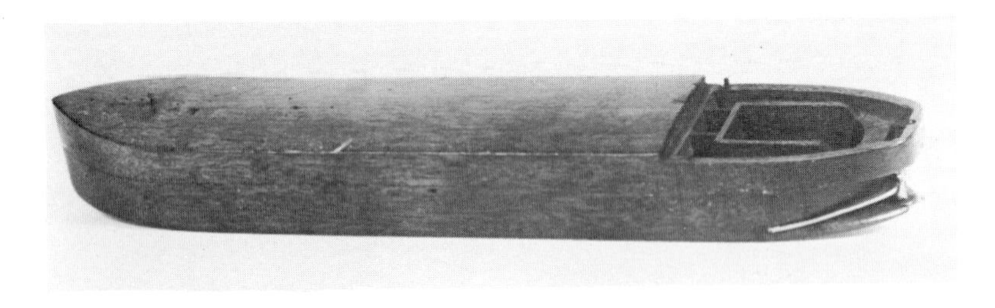

UNITED STATES PATENT OFFICE.

WILLIAM BAXTER AND WILLIAM BAXTER, JR., OF NEWARK, NEW JERSEY.

IMPROVEMENT IN STEAM CANAL-BOATS.

Specification forming part of Letters Patent No. 154,978, dated September 15, 1874; application filed August 25, 1874.

To all whom it may concern:

Be it known that we, WILLIAM BAXTER and WILLIAM BAXTER, Jr., both of the city of Newark, in the county of Essex and State of New Jersey, have invented an Improvement in Canal-Boats to be Propelled by Steam-Power, of which the following is a specification, reference being had to the accompanying drawings forming part thereof.

The object of our invention is the successful application of steam-power to the propulsion of boats on canals.

It is well known that such application of steam-power has been long regarded as very desirable, and it is equally well known that of the many attempts that have been made to accomplish it none have been so far successful as to effect the supersedure of horse-power for the purpose. The reasons of these repeated failures are to be found in the peculiar conditions and exigencies of canal navigation. These conditions and exigencies differ widely and essentially from those existing in navigation upon the broad waters of the ocean, or of our lakes and rivers. The canal-boat must be adapted to run within the narrow limits to which it is confined without causing injury to the canal structure. It must be prepared to encounter, without injury to its steam motive mechanism, frequent collisions with the banks of the canal, with the walls of locks, and with other boats, to which it is inevitably subjected. It must be propelled at a comparatively low rate of speed, which it may not exceed. It must have a certain minimum carrying capacity within the limits permitted by the depth of the canal and the size of the locks, and only a certain part of which capacity can be occupied by the motor, machinery, and fuel, as, if the prescribed limits in these respects are transcended, the cost of transportation by steam-power will exceed that by horse-power, and the former will prove economically a failure—a result as fatal as would be a mechanical failure; and the motor machinery, the fuel, and the freight must be so disposed and placed, relatively, that the boat shall be at all times perfectly trimmed, in order to use the utmost possible depth of the water without liability to touch ground. That the combining in a canal steamboat all the features necessary to meet these various conditions is a difficult achievement is proved by the fact that it has never been accomplished before.

We will now proceed to describe our canal-boat, in which are embodied the improvements whereby the results above referred to are secured; and it should be remarked that wherever we have departed from the form, construction, or arrangement commonly found in canal-boats, we have done so for a definite purpose, and to accomplish a predetermined result, as will be apparent from the description herein of our improvement.

These improvements relate, first, to the hull of the boat, and, in such relation, consist in combining a flat, plane bottom with sides the vertical section of which, in all their entire external surface, including bow and stern, presents vertical lines, at least below the water-line, or lines as nearly vertical as are the walls of the locks of the canal in which the boat is designed to run, thereby securing the greatest possible carrying capacity for the draft of the boat, the water, by the movement of the boat, being set in motion laterally only, leaving the water under the boat at rest, thereby obviating the stirring up of the bed of the canal, and obviating also the lifting of the bow, and consequent depression of the stern, when under way, which takes place with boats whose bottoms curve or incline upward at or near the bow; also, in giving to the bow such a form that its horizontal section at any point below the water-line presents a true Gothic arch—that is to say, the line on each side of the bow is the arc of a circle whose radius is at least equal to the breadth of the boat at the intersection of the bow with the sides, thereby securing the most advantageous adjustment possible of the maximum of carrying capacity to the minimum of resistance to the passage of the boat through the water; also, in giving to the stern, below the water-line, a form similar to the bow above indicated, with upper and lower guards for the protection of the propeller, formed by an overhanging portion of the hull around the stern above the water-line, and a lower guard for the same purpose extending in the plane of the boat's bottom underneath and beyond the propeller around the stern, the space

Many attempts were made throughout the canal states to find a method to link steam to canalboats. The Baxters were just one of the many people applying for a patent.

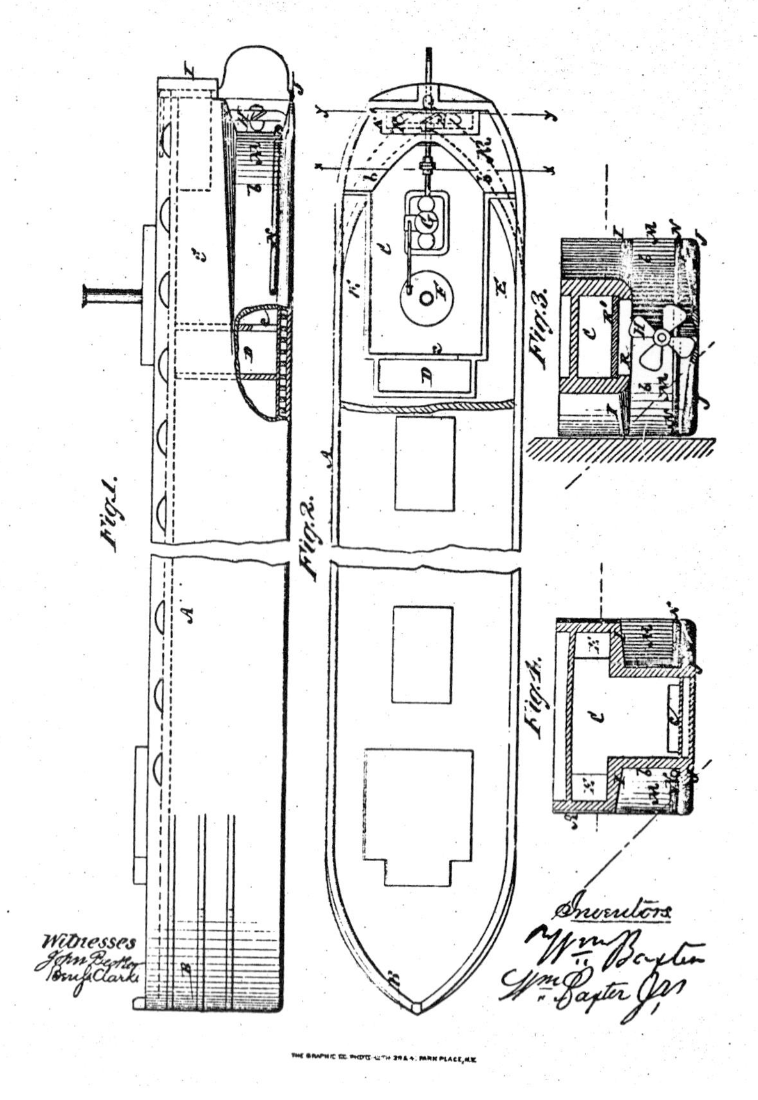

In addition to a model, each patent application had to be accompanied by a drawing. The Baxters had their drawing prepared by a New York professional.

(LEFT) Captain George Winters. (RIGHT) Daniel Winters, nephew of George Winters, cutting ice at the Port Murray boatyard.

EASTON EXPRESS–1913
Phillipsburg Section

ACTIVE OLD MAN IS HARD WORKER

Although Ninety-One Years of Age He Has No Idea of Retiring from his Employment

The above photograph is a perfect likeness of Captain George Winters, one of the oldest, if not the oldest resident of this town. Having passed the four score and eleven milestone of life, he is today as hale and hearty as any citizen born and bred in the county of Warren, who has reached that ripe old age.

Born in Karrsville, this county, in 1822, he set to work on the Morris Canal when but a lad of six years. When arriving at the age of twenty-four April 4, 1846, he married Miss Margaret

E. Thompson, a daughter of the long since deceased Sharps Thompson, who for years was the village blacksmith of that place. They were blessed with two children, Miss Mary Winters, with whom he resides at 224 South Main Street, and Benjamin Winters, of Boonton, both of whom are still living. Mr. Winters' wife died in 1900.

For fifty years he worked on the Morris Canal as a boatman, and on June 29, 1840, he was presented with a bible which he still has in his possession, and which bears the following inscription: "Presented to Captain George Winters, of the boat 'General Putman', weight–eight tons and five hundred pounds. June 29, 1840, by E. R. Biddle, Esq."

Mr. Winters, when asked yesterday, when he retired from active work, surprised the questioner by saying he had not given the question a thought as yet. He contents himself by working about the stable of Street Commissioner Jerome Durling, where he keeps the harness and horse in fine fettle.

Daniel P. Winters, aged 90, prepares for a 1926 unveiling of a plaque commemorating the Morris Canal placed on the Kresge Department Store in Newark.

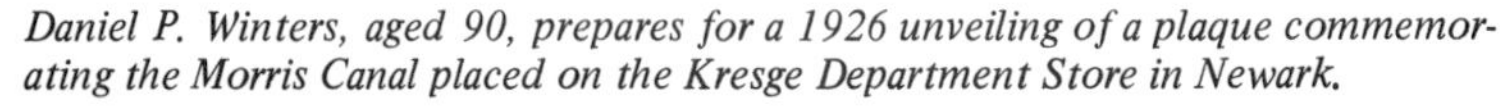

POWER HOUSE BURNED

Machinery, Building, and Tower at Plane No. 6 Destroyed on Sunday Night 1901

A fire that kept Port Colden in a state of excitement for several hours broke out on Sunday evening in the tower and powerhouse at Plane No. 6, near that place. The building and much of the machinery were entirely destroyed. The blaze was discovered about six o'clock shortly after a "hog" locomotive attached to a heavy train had passed, and the natural supposition is that a spark from the engine was the cause of the conflagration.

The plane at that point is a double one, one of three of that kind of the Morris Canal. For several years, however, only one track had been in use, but the heavy machinery necessary to operate both tracks was in place.

The flames had made but little headway when discovered, but the wooden building burned like timber and despite the efforts of the bucket brigade, made up of the young and old from Port Colden, the flames made headway rapidly. About 8:30 p.m., a heavy rain began to fall and this materially assisted the firemen, but before that time, much of the machinery contained in the building had been cracked and half the flume had been destroyed. The big drum used for lowering and raising boats, however, is thought to be uninjured.

Abram Scott, the Plane tender at No. 6, has been in charge there for thirty-three years and this is the first disaster of any kind that has occurred in all that time.

The building will probably be rebuilt at once, providing the Canal Closing Bill now in the Legislature fails to pass.

* * * * * * * *

Stewartsville was flooded when the Morris Canal gates were suddenly opened to prevent damage to the canal banks.

Hackettstown Gazette

1835–Silas S. Harvey built a large brewery along the banks of the Morris Canal. The brewery was of frame construction four stories high and had connected to it a large cove under the Morris Canal used as storage to keep the beer and ale cool. In 1859 the firm became Gardner Mitchel and Co. A few years later James Mitchel bought out other interest and operated it till 1890. In 1897 James Mitchel was operating a Canal store in North Hackettstown.

December 2, 1858–The banks of the Morris Canal gave way last night near Drakesville, carrying away a dwelling house in which were a woman and three children. They were all drowned.

June 20, 1874–Several parties from Newark embarked on one of the boats on the Morris Canal the other day for the purpose of taking a pleasure trip among the mountains. They passed through Hackettstown last evening, all apparently enjoying themselves. This is the second or third excursion of the kind this season, and it seems that the Morris Canal is growing in favor every year.

1874–The past season has been a dull one on the Morris Canal owing to the tedious delays in having the boats loaded.

January 2, 1875–A Mr. Cole of Washington skated from that place to Hackettstown one day last week, in 65 minutes, 15 of which were lost in crossing the planes; making a distance of 12 miles in 50 minutes. Who can beat this?

April 18, 1924–The eleven-mile level, that section of the Morris Canal from Plane 5, near Port Murray, to the guard lock three miles east of here, is practically dry. Last week Jacob Martenis, section foreman, received word to let out the water, and for several days gradually raised the gates so it would not overflow the banks of the waterways leading to the Musconetcong River. This eleven-mile level was one of the picturesque sections of the old canal and provided water for many farms for dairy purposes and fire protection, besides swimming, skating and fishing.

Chapter Twenty Four

Andy vs. Duke

As told to James Lee
by Patrick H. Burke of Phillipsburg

In the year 1879, Andrew Newman became proprietor of the Andover Hotel. This hotel lies on the east side along South Main Street in Phillipsburg, New Jersey, about one mile from the loading chutes at Port Delaware. On the opposite side of South Main Street lies the Morris Canal Basin, and the Andover Furnace Company wharves, where the boats tied up while they were unloading their cargo.

It was a very desirable place to tie up at because of its proximity to the town and also the Andover Hotel had a very large boat stable, which accommodated perhaps forty or fifty head of mules and horses. The service of this boat stable was very reasonable because all that the captains had to pay for was the feed and oats that their teams ate. Sheltering was free. This boat stable had a very wide space in the center, wide enough for two carriages to park side by side.

The hotel was very unique in itself. It was built in the earlier days of the Cooper regime, that famous pioneer iron master who was the founder and builder of the old Andover Iron Furnace.

Among the boat captains that usually tied up in the furnace canal basin was one by the name of John Merritt. He always went by the name of "Duke" Merritt. Duke was always considered the bully of the Morris Canal. He was a man that weighed about 185 pounds, stood about 5'10", broad shoulders and always carried a domination attitude, so much as that all the boat captains were afraid of him, as more than once they

---

Patrick H. Burke was born at Phillipsburg on June 22, 1870. He died at Phillipsburg on June 26, 1951. He was employed as a die maker.

saw his wicked punch drive a man clear across the towpath and into the canal.

Duke was always considered a man that possessed a good complement of fistic science, a little less than the then world champion Paddy Ryan. Duke also was known as having long fingers, and on several occasions had taken things that belonged to others, but no one would dare to accuse him of it, or if they would, they sure were in for a good sock on the jaw. The bully Duke was getting away with things very nicely.

Andrew Newman, the proprietor of the Andover Hotel, was very well known among the boat captains. They always called him "Andy" and Andy knew all the boat captains.

At closing up time of the barroom, Andy always had a habit of putting some money, " change," of about five dollars in one of the copper beer measures. This was done so the bartender or whoever would open up the barroom in the early mornings, would have some cash money to commence business with. The barroom always had a big early morning business rush with the canalers leaving port and the furnace men changing shifts, and particularly the furnace men, after making the morning's cast and ranking 60 or 70 tons of hot pig iron, refreshments were very desirable. Many times this money that was put in the copper beer measure was missing in the morning by the bartender when he opened up for business. Also, a few bottles of high grade liquors were missing each time along with the money. So mysterious was this incident that Andy was puzzled beyond all comprehension, as all the doors, windows and shutters were locked and securely fastened, so much so that Andy considered it impossible for anyone to enter the place without being seen or heard.

Andy assuming an attitude of "sign dumb" until one evening in the barroom, Andy overheard a whisper from one captain to another that Duke Merrit gave him a drink of good liquor at Number Eight Plane when he met Duke in the morning two days after Andy had missed his money and liquor.

Immediately the finger of suspicion began to point toward Duke, and Andy set out to solve the mystery. With only coal oil street lamps that were extinguished at midnight, it was somewhat difficult to recognize a man in utter darkness.

But anyhow, Andy decided to set the trap. He overheard a conversation two days before between Cidar Price and Beef Searles, two popular captains that ran the ditch. Cidar had told Beef that he passed Duke on the Seventeener as Cidar was running C R A Z Y liner he had no trouble in passing Duke with full speed ahead. Andy figured that if Cidar would reach furnace basin by five or six in the evening and on to Port Delaware, Duke would get through the last lock by 9:30 p.m., when all locks and planes would cease operating, and by Duke floating along slowly he would reach the furnace basin about 10:00 p.m., which would be the dark of the night. Andy, thinking all this over, went to work. The barroom was situated so that it could be entered from the outside by a front door off Main Street, and from a back door, which was well under the porch. It also could be entered on the inside by a door that led into a hallway. Here Andy decided to set the trap. He placed a dark lantern in the wall bracket so that it focused directly on the bar. Having lit the lantern and closing the slides which made the barroom 100% blackout, he closed the hallway door and

The Andover Hotel still can be found today along South Main Street in Phillipsburg.

laid in wait for whatever might turn up. Along about 2:00 a.m. in the still of the night, Andy heard a little noise at the back door of the barroom. The noise continued on a little longer, and Andy heard someone taking the money change out of the beer measure. On the impulse of the moment, Andy softly pushed the hallway door open and reached up to the dark lantern, he pulled the slide open. And to Andy's surprise the glare of the light shown full on Duke's surprised face.

"Duke, I have caught you at last," exclaimed Andy, as Duke stood bewildered holding a bottle of high grade liquor in one hand and his other hand in the pocket that contained the money that he just had taken from the beer measure. "You put back that bottle and the money you have in your pocket. You have gone as far as you are going to get with this racket. Duke, you will have to fight me for all that you have taken from here," exclaimed Andy. Duke quickly consented realizing that the sentence could be easily served. The condition of the fight was that the winner take all. This gave Duke hopes of still greater easy money. The date for the fight was set for three nights later. This gave Andy a chance to advertise the fight. He had printed place cards placed in the boat stable at Port Delaware, and in the saloons of the town, saying a prize fight will be held in the boat stable of the Andover Hotel on the mentioned date, between Duke Merritt and an unknown, admission $1.00, winner take all. This caused some anxiety among the local sports, knowing that the Duke was such a tough character, and that it would be a joke for any local man to attempt such a feat.

The night arrived that the big fight was to be pulled off. Coal oil lamps were arranged upon the straw posts so as to give as good a light as possible within an area of about ten feet square.

Among the notables that attended the fight was Robert (Bobby) Dempster, possessor of a string of thoroughbreds including "Iron King," a star of the race track. Theodore (Dory) Melter, Doc Bougardes and many others of race track fame.

Harry Rich, Edward (Humpy) Carpenter, Henry (Shanks) Acorn, and "Murf" Mooney, all skilled in the art of training and conditioning Bull Dogs, for long and ferocious battles.

Howard Carey, Richard (Dick) Gavin of game birds, (chickens) fame, both skilled in the art of fixing the steel gafts, that the drop of the spurs was so arranged that often in a main of cocks one or the other bird would be clipped off on the first fly. Joseph (Joe) Firth, later Mayor of the town and member of the New Jersey Assembly, Thomas (Slim) Kane, skilled in cutting the combs and gills off game roosters that was so essential for the protection of the bird from the opposing foe. Samuel B. Mutchler, D.W. (Wint) Hagerty, both later became members of the New Jersey Assembly. (Dock) Sheppard and Joe Hulshizer, both of live birds (pigeon) shooting matches fame, that were always known to kill 24 birds of a 25 bird shoot, and many other of the sporting fraternity were on hand to see the fracus.

With the stage all set and anxiety registering about 95 in the shade, Duke Merritt appeared in the arena, followed by the unknown in person—no less than the tall lanky proprietor of the Andover Hotel, Andy Newman. Both stripped to the waist, and trousers rolled up half way to the knees, the finishing touches were in order.

Joe Firth volunteered to set as referree and the rules that governed the fight were somewhat similar to the London Prize Ring Rules. Continuous, no rounds, no clinching, no biting, and no kicking when a man is down, and bare fists.

It was just 12 o'clock midnight when the referree gave the word go," and both men let go. Fast and furious were their blows, with their straights from the shoulders, cross counters, hooks and jabs. Each man scored two knock downs in the first ten minutes, and bare fists and faces began to take on a crimson appearance. After that, the men began to get a little bit groggy. Then Andy scored on the knock down. And Duke had much more difficulty in getting on his feet. The time Duke took to get back on his feet gave Andy a new fresh spirit, and with a Paddy Ryan brand of scientific fighting, Andy sent his blows straight to Duke's jaw, with such force that he put Duke dazed reeling and finally collapsing. He dropped to the ground down and out.

Immediately referree Firth jumped into the arena, and declared Andy the winner, and with the decision went about sixty dollars of gate receipts. The fight lasted just fifteen minutes by Bobby Dempster's stop watch. The fans were both elated and very well satisfied with the beautiful pugilistic exhibition of cleverness, skill and tact, that both men displayed during the fight from start to finish.

The next day rumors and gossip in a sportsmanship way were plenty, and Duke openly and honestly declared that Andy was the better man, and that he was through with the fighting spirit in the days to come. As Port Delaware was a terminal point, many boats and their captains laid over here for the weekend. During Saturday evening, Sunday and Sunday evening there were a variety of programs scheduled for the enjoyment and benefit of the men who were engaged in the transportation of anthracite.

One very popular program was held on Sunday about 10 a.m. and was in the form of revival services. The sweet melodies that were rendered at these services by local talent with guitars and banjos were the admiration by all that spent the weekend at this shipping resort. The most noted speaker at these revivals was Mr. John (Scotty) Johnson, who always had for his subject "The Power of Habit (Temperance)." This appeared to have interested Duke Merritt so much that he finally yielded to impulse and became converted to the routine of the revival meeting. After that, Duke would exercise special effort to attend that Sunday afternoon program. As he began to see life in a more pleasanter aspect, his endurance earned for him the admiration and respect of his fellow Captains. This is one of the many incidents that occurred during the ninety years of existence of this waterway. And as Duke would often admit that the defeat he received from Andy was only "brushing the clouds away" to enable Duke to "see the dawn of a bright and better day."

INDEX